FRONTIERS IN
MASSIVE
DATA
ANALYSIS

Committee on the Analysis of Massive Data

Committee on Applied and Theoretical Statistics

Board on Mathematical Sciences and Their Applications

Division on Engineering and Physical Sciences

NATIONAL RESEARCH COUNCIL
OF THE NATIONAL ACADEMIES

THE NATIONAL ACADEMIES PRESS
Washington, D.C.
www.nap.edu

THE NATIONAL ACADEMIES PRESS   500 Fifth Street, NW   Washington, DC 20001

NOTICE: The project that is the subject of this report was approved by the Governing Board of the National Research Council, whose members are drawn from the councils of the National Academy of Sciences, the National Academy of Engineering, and the Institute of Medicine. The members of the committee responsible for the report were chosen for their special competences and with regard for appropriate balance.

This project was supported by the National Security Agency under contract number NSA H98230-09-C-0407. Any opinions, findings, conclusions, or recommendations expressed in this publication are those of the author(s) and do not necessarily reflect the views of the organizations or agencies that provided support for the project.

International Standard Book Number 13:   978-0-309-28778-4
International Standard Book Number 10:   0-309-28778-2
Library of Congress Control Number: 2013944743

Cover: Image courtesy of Jonathan Bachrach, University of California, Berkeley.

Additional copies of this report are available from the National Academies Press, 500 Fifth Street, NW, Keck 360, Washington, DC 20001; (800) 624-6242 or (202) 334-3313; http://www.nap.edu.

Suggested citation: National Research Council. 2013. *Frontiers in Massive Data Analysis*. Washington, D.C.: The National Academies Press.

THE NATIONAL ACADEMIES
Advisers to the Nation on Science, Engineering, and Medicine

The **National Academy of Sciences** is a private, nonprofit, self-perpetuating society of distinguished scholars engaged in scientific and engineering research, dedicated to the furtherance of science and technology and to their use for the general welfare. Upon the authority of the charter granted to it by the Congress in 1863, the Academy has a mandate that requires it to advise the federal government on scientific and technical matters. Dr. Ralph J. Cicerone is president of the National Academy of Sciences.

The **National Academy of Engineering** was established in 1964, under the charter of the National Academy of Sciences, as a parallel organization of outstanding engineers. It is autonomous in its administration and in the selection of its members, sharing with the National Academy of Sciences the responsibility for advising the federal government. The National Academy of Engineering also sponsors engineering programs aimed at meeting national needs, encourages education and research, and recognizes the superior achievements of engineers. Dr. C. D. Mote, Jr., is president of the National Academy of Engineering.

The **Institute of Medicine** was established in 1970 by the National Academy of Sciences to secure the services of eminent members of appropriate professions in the examination of policy matters pertaining to the health of the public. The Institute acts under the responsibility given to the National Academy of Sciences by its congressional charter to be an adviser to the federal government and, upon its own initiative, to identify issues of medical care, research, and education. Dr. Harvey V. Fineberg is president of the Institute of Medicine.

The **National Research Council** was organized by the National Academy of Sciences in 1916 to associate the broad community of science and technology with the Academy's purposes of furthering knowledge and advising the federal government. Functioning in accordance with general policies determined by the Academy, the Council has become the principal operating agency of both the National Academy of Sciences and the National Academy of Engineering in providing services to the government, the public, and the scientific and engineering communities. The Council is administered jointly by both Academies and the Institute of Medicine. Dr. Ralph J. Cicerone and Dr. C. D. Mote, Jr., are chair and vice chair, respectively, of the National Research Council.

www.national-academies.org

Acknowledgments

This report has been reviewed in draft form by individuals chosen for their diverse perspectives and technical expertise, in accordance with procedures approved by the National Research Council's Report Review Committee. The purpose of this independent review is to provide candid and critical comments that will assist the institution in making its published report as sound as possible and to ensure that the report meets institutional standards for objectivity, evidence, and responsiveness to the study charge. The review comments and draft manuscript remain confidential to protect the integrity of the deliberative process. We wish to thank the following individuals for their review of this report:

Amy Braverman, Jet Propulsion Laboratory,
John Bruning, Corning Tropel Corporation (retired),
Jeffrey Hammerbacher, Cloudera,
Iain Johnstone, Stanford University,
Larry Lake, University of Texas,
Richard Sites, Google, Inc., and
Hal Stern, University of California, Irvine.

Although the reviewers listed above have provided many constructive comments and suggestions, they were not asked to endorse the conclusions or recommendations nor did they see the final draft of the report before its release. The review of this report was overseen by Michael Goodchild of the University of California, Santa Barbara. Appointed by the National Research Council, he was responsible for making certain that an independent examination of this report was carried out in accordance with institutional procedures and that all review comments were carefully considered. Responsibility for the final content of this report rests entirely with the authoring committee and the institution.

The committee also acknowledges the valuable contribution of the following individuals, who provided input at the meetings on which this report is based or through other communications:

Léon Bottou, NEC Laboratories,
Jeffrey Dean, Google, Inc.,
John Gilbert, University of California, Santa Barbara,
Jeffrey Hammerbacher, Cloudera,
Patrick Hanrahan, Stanford University,
S. Muthu Muthukrishnan, Rutgers, The State University of New Jersey,
Ben Shneiderman, University of Maryland,
Michael Stonebraker, Massachusetts Institute of Technology, and
J. Anthony Tyson, University of California, Davis.

Contents

Summary

THE PROMISE AND PERILS OF MASSIVE DATA

Experiments, observations, and numerical simulations in many areas of science and business are currently generating terabytes of data, and in some cases are on the verge of generating petabytes and beyond. Analyses of the information contained in these data sets have already led to major breakthroughs in fields ranging from genomics to astronomy and high-energy physics and to the development of new information-based industries. Traditional methods of analysis have been based largely on the assumption that analysts can work with data within the confines of their own computing environment, but the growth of "big data" is changing that paradigm, especially in cases in which massive amounts of data are distributed across locations.

While the scientific community and the defense enterprise have long been leaders in generating and using large data sets, the emergence of e-commerce and massive search engines has led other sectors to confront the challenges of massive data. For example, Google, Yahoo!, Microsoft, and other Internet-based companies have data that is measured in exabytes (10^{18} bytes). Social media (e.g., Facebook, YouTube, Twitter) have exploded beyond anyone's wildest imagination, and today some of these companies have hundreds of millions of users. Data mining of these massive data sets is transforming the way we think about crisis response, marketing, entertainment, cybersecurity, and national intelligence. It is also transforming how we think about information storage and retrieval. Collections of documents, images, videos, and networks are being thought of not merely as bit strings to be stored, indexed, and retrieved, but also as potential sources of discovery and knowledge, requiring sophisticated analysis techniques that go far beyond classical indexing and keyword counting, aiming to find relational and semantic interpretations of the phenomena underlying the data.

A number of challenges in both data management and data analysis require new approaches to support the big data era. These challenges span generation of the data, preparation for analysis, and policy-related challenges in its sharing and use, including the following:

- Dealing with highly distributed data sources,
- Tracking data provenance, from data generation through data preparation,
- Validating data,
- Coping with sampling biases and heterogeneity,
- Working with different data formats and structures,
- Developing algorithms that exploit parallel and distributed architectures,
- Ensuring data integrity,
- Ensuring data security,
- Enabling data discovery and integration,
- Enabling data sharing,
- Developing methods for visualizing massive data,
- Developing scalable and incremental algorithms, and
- Coping with the need for real-time analysis and decision-making.

To the extent that massive data can be exploited effectively, the hope is that science will extend its reach, and technology will become more adaptive, personalized, and robust. It is appealing to imagine, for example, a health-care system in which increasingly detailed data are maintained for each individual—including genomic, cellular, and environmental data—and in which such data can be combined with data from other individuals and with results from fundamental biological and medical research so that optimized treatments can be designed for each individual. One can also envision numerous business opportunities that combine knowledge of preferences and needs at the level of single individuals with fine-grained descriptions of goods, skills, and services to create new markets.

It is natural to be optimistic about the prospects. Several decades of research and development in databases and search engines have yielded a wealth of relevant experience in the design of scalable data-centric technology. In particular, these fields have fueled the advent of cloud computing and other parallel and distributed platforms that seem well suited to massive data analysis. Moreover, innovations in the fields of machine learning, data mining, statistics, and the theory of algorithms have yielded data-analysis methods that can be applied to ever-larger data sets. However, such optimism must be tempered by an understanding of the major difficulties that arise in attempting to achieve the envisioned goals. In part, these difficulties are those familiar from implementations of large-scale databases—finding and mitigating bottlenecks, achieving simplicity and generality of the programming interface, propagating metadata, designing a system that is robust to hardware failure, and exploiting parallel and distributed hardware—all at an unprecedented scale. But the challenges for massive data go beyond the storage, indexing, and querying that have been the province of classical database systems (and classical search engines) and, instead, hinge on the ambitious goal of *inference*. Inference is the problem of turning data into knowledge, where knowledge often is expressed in terms of entities that are not present in the data per se but are present in models that one uses to interpret the data. Statistical rigor is necessary to justify the inferential leap from data to knowledge, and many difficulties arise in attempting to bring statistical principles to bear on massive data. Overlooking this foundation may yield results that are not useful at best, or harmful at worst. In any discussion of massive data and inference, it is essential to be aware that it is quite possible to turn data into something resembling knowledge when actually it is not. Moreover, it can be quite difficult to know that this has happened.

Indeed, many issues impinge on the quality of inference. A major one is that of "sampling bias." Data may have been collected according to a certain criterion (for example, in a way that favors "larger" items over "smaller" items), but the inferences and decisions made may refer to a different sampling criterion. This issue seems likely to be particularly severe in many massive data sets, which often consist of many subcollections of data, each collected according to a particular choice of sampling criterion and with little control over the overall composition. Another major issue is "provenance." Many systems involve layers of inference, where "data" are not the original observations but are the products of an inferential procedure of some kind. This often occurs, for example, when there are missing entries in the original data. In a large system involving interconnected inferences, it can be difficult to avoid circularity, which can introduce additional biases and can amplify noise. Finally, there is the major issue of controlling error rates when many hypotheses are being considered. Indeed, massive data sets generally involve growth not merely in the number of individuals represented (the "rows" of the database) but also in the number of descriptors of those individuals (the "columns" of the database). Moreover, we are often interested in the predictive ability associated with combinations of the descriptors; this can lead to exponential growth in the number of hypotheses considered, with severe consequences for error rates. That is, a naive appeal to a "law of large numbers" for massive data is unlikely to be justified; if anything, the perils associated with statistical fluctuations may actually *increase* as data sets grow in size.

While the field of statistics has developed tools that can address such issues in principle, in the context of massive data care must be taken with all such tools for two main reasons: (1) all statistical tools are based on assumptions about characteristics of the data set and the way it was sampled, and those assumptions may be violated in the process of assembling massive data sets; and (2) tools for assessing errors of procedures, and for diagnostics, are themselves computational procedures that may be computationally infeasible as data sets move into the massive scale.

In spite of the cautions raised above, the Committee on the Analysis of Massive Data believes that many of the challenges involved in performing inference on massive data can be confronted usefully. These challenges must be addressed through a major, sustained research effort that is based solidly on both inferential and computational principles. This research effort must develop scalable computational infrastructures that embody inferential principles that themselves are based on considerations of scale. The research must take account of real-time decision cycles and the management of trade-offs between speed and accuracy. And new tools are needed to bring humans into the data-analysis loop at all stages, recognizing that knowledge is often subjective and context-dependent and that some aspects of human intelligence will not be replaced anytime soon by machines.

The current report is the result of a study that addressed the following charge:

- Assess the current state of data analysis for mining of massive sets and streams of data,
- Identify gaps in current practice and theory, and
- Propose a research agenda to fill those gaps.

Thus, this report examines the frontiers of research that is enabling the analysis of massive data. The major research areas covered are as follows:

- Data representation, including characterizations of the raw data and transformations that are often applied to data, particularly transformations that attempt to reduce the representational complexity of the data;
- Computational complexity issues and how the understanding of such issues supports characterization of the computational resources needed and of trade-offs among resources;
- Statistical model-building in the massive data setting, including data cleansing and validation;

- Sampling, both as part of the data-gathering process but also as a key methodology for data reduction; and
- Methods for including humans in the data-analysis loop through means such as crowdsourcing, where humans are used as a source of training data for learning algorithms, and visualization, which not only helps humans understand the output of an analysis but also provides human input into model revision.

CONCLUSIONS

The research and development necessary for the analysis of massive data goes well beyond the province of a single discipline, and one of the main conclusions of this report is the need for a thoroughgoing interdisciplinarity in approaching problems of massive data. Computer scientists involved in building big-data systems must develop a deeper awareness of inferential issues, while statisticians must concern themselves with scalability, algorithmic issues, and real-time decision-making. Mathematicians also have important roles to play, because areas such as applied linear algebra and optimization theory (already contributing to large-scale data analysis) are likely to continue to grow in importance. Also, as just mentioned, the role of human judgment in massive data analysis is essential, and contributions are needed from social scientists and psychologists as well as experts in visualization. Finally, domain scientists and users of technology have an essential role to play in the design of any system for data analysis, and particularly so in the realm of massive data, because of the explosion of design decisions and possible directions that analyses can follow.

The current report focuses on the technical issues—computational and inferential—that surround massive data, consciously setting aside major issues in areas such as public policy, law, and ethics that are beyond the current scope.

The committee reached the following conclusions:

- Recent years have seen rapid growth in parallel and distributed computing systems, developed in large part to serve as the backbone of the modern Internet-based information ecosystem. These systems have fueled search engines, electronic commerce, social networks, and online entertainment, and they provide the platform on which massive data analysis issues have come to the fore. Part of the challenge going forward is the problem of scaling these systems and algorithms to ever-larger collections of data. It is important to acknowledge, however, that the goals of massive data analysis go beyond the computational and representational issues that have been province of classical search engines and database processing to tackling the challenges of statistical inference, where the goal is to turn data into knowledge and to support effective decision-making. Assertions of knowledge require control over errors, and a major part of the challenge of massive data analysis is that of developing statistically well-founded procedures that provide control over errors in the setting of massive data, recognizing that these procedures are themselves computational procedures that consume resources.

- There are many sources of potential error in massive data analysis, many of which are due to the interest in "long tails" that often accompany the collection of massive data. Events in the "long tail" may be vanishingly rare, even in a massive data set. For example, in consumer-facing information technology, where the goal is increasingly that of providing fine-grained, personalized services, there may be little data available for many individuals, even in very large data sets. In science, the goal is often that of finding unusual or rare phenomena, and evidence for such phenomena may be weak, particularly when one considers the increase in error rates associated with searching over large classes of hypotheses. Other sources of error that are prevalent in massive data include the high-dimensional nature of many data sets, issues of heterogeneity, biases arising from uncontrolled sampling patterns, and unknown provenance of items in a database. In general, data analysis is based on assumptions, and the assumptions underlying many classical data analysis methods are likely to be broken in massive data sets.

- Massive data analysis is not the province of any one field, but is rather a thoroughly interdisciplinary enterprise. Solutions to massive data problems will require an intimate blending of ideas from computer science and statistics, with essential contributions also needed from applied and pure mathematics, from optimization theory, and from various engineering areas, notably signal processing and information theory. Domain scientists and users of technology also need to be engaged throughout the process of designing systems for massive data analysis. There are also many issues surrounding massive data (most notably privacy issues) that will require input from legal scholars, economists, and other social scientists, although these aspects are not covered in the current report. In general, by bringing interdisciplinary perspectives to bear on massive data analysis, it will be possible to discuss trade-offs that arise when one jointly considers the computational, statistical, scientific, and human-centric constraints that frame a problem. When considering parts of the problem in isolation, one may end up trying to solve a problem that is more general than is required, and there may be no feasible solution to that broader problem; a suitable cross-disciplinary outlook can point researchers toward an essential refocusing. For example, absent appropriate insight, one might be led to analyzing worst-case algorithmic behavior, which can be very difficult or misleading, whereas a look at the totality of a problem could reveal that average-case algorithmic behavior is quite appropriate from a statistical perspective. Similarly, knowledge of typical query generation might allow one to confine an analysis to a relatively simple subset of all possible queries that would have to be considered in a more general case. And the difficulty of parallel programming in the most general settings may be sidestepped by focusing on useful classes of statistical algorithms that can be implemented with a simplified set of parallel programming motifs; moreover, these motifs may suggest natural patterns of storage and access of data on distributed hardware platforms.

- While there are many sources of data that are currently fueling the rapid growth in data volume, a few forms of data create particularly interesting challenges. First, much current data involves human language and speech, and increasingly the goal with such data is to extract aspects of the semantic meaning underlying the data. Examples include sentiment analysis, topic models of documents, relational modeling, and the full-blown semantic analyses required by question-answering systems. Second, video and image data are increasingly prevalent, creating a range of challenges in large-scale compression, image processing, computational vision, and semantic analysis. Third, data are increasingly labeled with geo-spatial and temporal tags, creating challenges in maintaining coherence across spatial scales and time. Fourth, many data sets involve networks and graphs, with inferential questions hinging on semantically rich notions such as "centrality" and "influence." The deeper analyses required by data sources such as these involve difficult and unsolved problems in artificial intelligence and the mathematical sciences that go beyond near-term issues of scaling existing algorithms. The committee notes, however, that massive data itself can provide new leverage on such problems, with machine translation of natural language a frequently cited example.

- Massive data analysis creates new challenges at the interface between humans and computers. As just alluded to, many data sets require semantic understanding that is currently beyond the reach of algorithmic approaches and for which human input is needed. This input may be obtained from the data analyst, whose judgment is needed throughout the data analysis process, from the framing of hypotheses to the management of trade-offs (e.g., errors versus time) to the selection of questions to pursue further. It may also be obtained from crowdsourcing, a potentially powerful source of inputs that must be used with care, given the many kinds of errors and biases that can arise. In either case, there are many challenges that need to be faced in the design of effective visualizations and interfaces and, more generally, in linking human judgment with data analysis algorithms.

- Many data sources operate in real time, producing data streams that can overwhelm data analysis pipelines. Moreover, there is often a desire to make decisions rapidly, perhaps also in real time. These temporal issues provide a particularly clear example of the need for further dialog between statistical and computational researchers. Statistical research has rarely considered constraints due to real-time decision-making in the development of data analysis algorithms, and computational research has rarely considered the computational complexity of algorithms for managing statistical risk.

- There is a major need for the development of "middleware"—software components that link high-level data analysis specifications with low-level distributed systems architectures. Much of the work on these software components can borrow from tools already developed in scientific computing instances, but the focus will need to change, with algorithmic solutions constrained by statistical needs. There is also a major need for software targeted to end users, such that relatively naive users can carry out massive data analysis without a full understanding of the underlying systems issues and statistical issues. However, this is not to suggest that the end goal of massive data analysis software is to develop turnkey solutions. The exercise of effective human judgment will always be required in data analysis, and this judgment needs to be based on an understanding of statistics and computation. The development of massive data analysis systems needs to proceed in parallel with a major effort to educate students and the workforce in statistical thinking and computational thinking.

As part of the study that led to this report, the Committee on the Analysis of Massive Data developed a taxonomy of some of the major algorithmic problems arising in massive data analysis. It is hoped that that this proposed taxonomy might help organize the research landscape and also provide a point of departure for the design of the middleware called for above. This taxonomy identifies major tasks that have proved useful in data analysis, grouping them roughly according to mathematical structure and computational strategy. Given the vast scope of the problem of data analysis and the lack of existing general-purpose computational systems for massive data analysis from which to generalize, there may certainly be other ways to cluster these computational tasks, and the committee intends this list only to initiate a discussion. The committee identified the following seven major tasks:

1. Basic statistics,
2. Generalized N-body problem,
3. Graph-theoretic computations,
4. Linear algebraic computations,
5. Optimization,
6. Integration, and
7. Alignment problems.

For each of these computational classes, there are computational constraints that arise within any particular problem domain that help to determine the specialized algorithmic strategy to be employed. Most work in the past has focused on a setting that involves a single processor with the entire data set fitting in random access memory (RAM). Additional important settings for which algorithms are needed include the following:

- The streaming setting, in which data arrive in quick succession, and only a subset can be stored;
- The disk-based setting, in which the data are too large to store in RAM but fit on one machine's disk;
- The distributed setting, in which the data are distributed over multiple machines' RAMs or disks; and
- The multi-threaded setting, in which the data lie on one machine having multiple processors that share RAM.

Training students to work in massive data analysis will require experience with massive data and with computational infrastructure that permits the real problems associated with massive data to be revealed. The availability of benchmarks, repositories (of data and software), and computational infrastructure will be a necessity in training the next generation of "data scientists." The same point, of course, can be made for academic research: significant new ideas will only emerge if academics are exposed to real-world massive data problems.

Finally, the committee emphasizes that massive data analysis is not one problem or one methodology. Data are often heterogeneous, and the best attack on a problem may involve finding sub-problems, where the best solution may be chosen for computational, inferential, or interpretational reasons. The discovery of such sub-problems might itself be an inferential problem. On the other hand, data often provide partial views onto a problem, and the solution may involve fusing multiple data sources. These perspectives of segmentation versus fusion will not be in conflict often, but substantial thought and domain knowledge may be required to reveal the appropriate combination.

One might hope that general, standardized procedures might emerge that can be used as a default for any massive data set, in much the way that the Fast Fourier Transform is a default procedure in classical signal processing. However, the committee is pessimistic that such procedures exist in general. That is not to say that useful general procedures and pipelines will not emerge; indeed, one of the goals of this report has been to suggest approaches for designing such procedures. But it is important to emphasize the need for flexibility and for tools that are sensitive to the overall goals of an analysis; massive data analysis cannot, in general, be reduced to turnkey procedures that consumers can use without thought. Rather, the design of a system for massive data analysis will require engineering skill and judgment, and deployment of such a system will require modeling decisions, skill with approximations, attention to diagnostics, and robustness. As much as the committee expects to see the emergence of new software and hardware platforms geared to massive data analysis, it also expects to see the emergence of a new class of engineers whose skill is the management of such platforms in the context of the solution of real-world problems.

1

Introduction

THE CHALLENGE

Although humans have gathered data since the beginning of recorded history—indeed, data gathered by ancestral humans provides much of the raw material for the reconstruction of human history—the rate of acquisition of data has surged in recent years, with no end in sight. Expectations have surged as well, with hopes for new scientific discoveries pinned on emerging massive collections of biological, physical, and social data, and with major areas of the economy focused on the commercial implications of massive data.

Although it is difficult to characterize all of the diverse reasons for the rapid growth in data, a few factors are worth noting. First, many areas of science are in possession of mature theories that explain a wide range of phenomena, such that further testing and elaboration of these theories requires probing extreme phenomena. These probes often generate very large data sets. An example is the world of particle physics, where massive data (e.g., petabytes per year for the Large Hadron Collider; 1 petabyte is 10^{15} bytes) arises from the new accelerators designed to test aspects of the Standard Model of particle physics. Second, many areas of science and engineering have become increasingly exploratory, with large data sets being gathered outside the context of any particular theory in the hope that new phenomena will emerge. Examples include the massive data arising from genome sequencing projects (which can accumulate terabytes (10^{12} bytes) of data for each project) as well as the massive data expected to arise from the Large Synoptic Survey Telescope, which will be measured in petabytes.

Rapid advances in cost-effective sensing mean that engineers can readily collect massive amounts of data about complex systems, such as those for communication networks, the electric grid, and transportation and financial systems, and use that data for management and control. Third, much human activity now takes place on the Internet, and this activity generates data that has substantial commercial and scientific value. In particular, many commercial enterprises are aiming to provide personalized services that adapt to individual behaviors and preferences as revealed by data associated with the individual. Fourth, connecting these other trends is the significant growth in the deployment of sensor networks that record biological, physical, and social phenomena at ever-increasing scale, and these sensor networks are increasingly interconnected.

In general, the hope is that if massive data could be exploited effectively, science would extend its reach, and technology would become more adaptive, personalized, and robust. It is appealing to imagine, for example, a health-care system in which increasingly detailed data are maintained for each individual—including genomic, cellular, and environmental data—and in which such data can be combined with data from other individuals and with results from fundamental biological and medical research, so that optimized treatments can be designed for each individual. One can also envision numerous microeconomic consequences of massive data analysis where preferences and needs at the level of single individuals are combined with fine-grained descriptions of goods, skills, and services to create new markets. In general, what is particularly notable about the recent rise in the prevalence of "big data" is not merely the size of modern data sets, but rather that their fine-grained nature permits inferences and decisions at the level of single individuals.

It is natural to be optimistic about the prospects. Several decades of research and development in databases and search engines have yielded a wealth of relevant experience in the design of scalable data-centric technology. In particular, these fields have fueled the advent of cloud computing and other parallel and distributed platforms that seem well suited to massive data analysis. Moreover, innovations in the fields of machine learning, data mining, statistics, and the theory of algorithms have yielded data-analysis methods that can be applied to ever-larger data sets. When combined with arguments that simple algorithms can work better than more sophisticated algorithms on large-scale data (see, e.g., Halevy et al., 2009), it is natural to be bullish on big data.[1]

While not entirely unwarranted, such optimism overlooks a number of major difficulties that arise in attempting to achieve the goals that

---

[1] This report uses the terms "big data" and "massive data" interchangeably to refer to data at massive scale.

are envisioned in discussions of massive data. In part these difficulties are those familiar from implementations of large-scale databases—involving finding and mitigating bottlenecks, achieving simplicity and generality of the programming interface, propagating metadata, designing a system that is robust to hardware failure, and exploiting parallel and distributed hardware—all at an unprecedented scale. But the goals for massive data go beyond the storage, indexing, and querying that have been the province of classical database systems (and classical search engines), instead focusing on the ambitious goal of *inference*. Inference is the problem of turning data into knowledge, where knowledge often is expressed in terms of variables (e.g., a patient's general state of health, or a shopper's tendency to buy) that are not present in the data per se, but are present in models that one uses to interpret the data. Statistical principles are needed to justify the inferential leap from data to knowledge, and many difficulties arise in attempting to bring these principles to bear on massive data. Operating in the absence of these principles may yield results that are not useful at best or harmful at worst. In any discussion of massive data and inference, it is essential to be aware that it is quite possible to turn data into something resembling knowledge but which actually is not. Moreover, it can be quite difficult to know that this has happened.

Consider a database where the rows correspond to people and the columns correspond to "features" that are used to describe people. If the database contains data on only a thousand people, it may suffice to measure only a few dozen features (e.g., age, gender, years of education, city of residence) to make the kinds of distinctions that may be needed to support assertions of "knowledge." If the database contains data on several billion people, however, we are likely to have heightened expectations for the data, and we will want to measure many more features (e.g., latest magazine read, culinary preferences, genomic markers, travel patterns) to support the wider range of inferences that we wish to make on the basis of the data. We might roughly imagine the number of features scaling linearly in the number of individuals. Now, the knowledge we wish to obtain from such data is often expressed in terms of combinations of the features. For example, if one lives in Memphis, is a male, enjoys reading about gardening, and often travels to Japan, what is the probability that the person will click on an ad about life insurance? The problem is that there are exponential numbers of such combinations of features and, in any given data set, a vast number of these combinations will appear to be highly predictive of any given outcome by chance alone.

As this scenario suggests, a naive appeal to a "law of large numbers" for massive data is unlikely to be justified. If anything, we should expect the perils associated with statistical fluctuations to *increase* as data sets grow in size. Of course, if we do not ask new questions as the data grow in size, but are content to more precisely answer old questions, then statistical error rates may not grow as the data scale. But that is not the perspective that underlies the current interest in massive data.

The field of statistics aims to provide a mathematical understanding of inference, quantifying the degree of support that data offer for assertions of knowledge as well as providing a basis for evaluating actions that are proposed on the basis of these assertions. The field has developed tools not only for computing estimates and evaluating hypotheses, but also for assessing the error rates of such procedures. One example of such a tool is "cross-validation," whereby one holds out a certain fraction of the data (the "held-out data"), runs the estimation procedure on the rest of the data (the "training data"), and tests on the held-out data. This accords well with the intuitive notion that an assertion can be viewed as "knowledge" if it applies not merely to the data at hand, but also to additional data. A difficulty, however, is that if one runs such a procedure many times on a fixed set of held-out data, for example, with many combinations of features, then many combinations will appear to be highly supported by both the training data and the held-out data, again by chance alone.

There are many additional issues that impinge on the quality of inference. A major issue is that of sampling bias. Data may have been collected according to a certain criterion (for example, in a way that favors "larger" items over "smaller" items), but the inferences and decisions we wish to make may refer to a different sampling criterion. This issue seems likely to be particularly severe in many massive data sets, which often consist of many subcollections of data, each collected according to a particular choice of sampling criterion and with little control over the overall composition. Another major issue is provenance. Many systems involve layers of inference, where "data" are not the original observations but are the products of an inferential procedure of some kind. This often occurs, for example, when there are missing entries in the original data. In a large system involving interconnected inferences, it can be difficult to avoid circularity. Circularity can introduce additional biases and amplify noise.

Many of these issues can be addressed in principle. For example, there are sophisticated statistical tools that can assess the errors in error-assessment procedures. But in the context of massive data, care must be taken with all such tools, for two main reasons:

1. These tools are all based on assumptions and they can fail when the assumptions are not met. Such assumptions include various assertions regarding sampling, stationarity, independence, and so on. Unfortunately, massive data sets are often collected in ways that seem most likely to break these assumptions.

2. Tools for assessing errors of procedures and for diagnostics are themselves computational procedures that make demands on the computing infrastructure. These demands may be infeasible for massive data sets, even when the underlying procedure is feasible.

Having aimed to temper some of the optimism that is often found in contemporary discussions of massive data, the committee does not want to align itself with an unduly pessimistic stance. The committee believes that many of the issues involved in performing inference on massive data can be confronted usefully, giving rise to an engineering discipline that is based solidly on both inferential and computational principles. But this will necessitate a major, sustained research effort that will require due attention to both the opportunities and the perils of massive data. It is necessary to develop scalable computational infrastructures that embody inferential principles that themselves are based on considerations of scale. Researchers will need to worry about real-time decision cycles and the management of trade-offs between speed and accuracy. While inference is increasingly used to power "data products" that are generated by machines—advanced search engines, movie recommender systems, news story and advertisement selection, and so on—there is also a need to develop tools for bringing humans into the data-analysis loop at all stages, because knowledge is often subjective and context-dependent, and there are aspects of human intelligence that (for the foreseeable future) are beyond the capability of machines.

This effort goes well beyond the province of a single discipline, and one of the main conclusions of this report is the need for a thoroughgoing interdisciplinarity in approaching problems of massive data. The major roles that computer scientists and statisticians have to play have already been alluded to above, and the committee emphasizes that the computer scientists involved in building big data systems must develop a deeper awareness of inferential issues, while statisticians must concern themselves with scalability, algorithmic issues, and real-time decision-making. Mathematicians also have important roles to play, with areas such as applied linear algebra already contributing to large-scale data analysis and likely to continue to grow in importance. But while the focus in much applied mathematical work has historically been on the control of small numerical errors, in massive data analysis, small numerical errors are likely to be dominated by statistical fluctuations and biases, and new paradigms need to be considered. This report also highlights the transdisciplinary domain of optimization theory, which already plays a major role in modern data analysis, but which also needs further shaping so as to meet the particular context of massive data. Also, as mentioned above, the role of human judgment in massive data analysis is essential, and contributions are needed from social scientists and psychologists as well as experts in visualization.

Finally, domain scientists and users of technology also have an essential role to play in the design of any system for data analysis, and particularly so in the realm of massive data, with the explosion of design decisions and possible directions that analyses can follow.

The focus in this report is on the technical issues—computational and inferential—that surround massive data. The committee recognizes that this focus is restrictive; there are also major issues to consider in areas such as public policy, law, and ethics. In particular, issues of privacy and the ownership of data are of major concern, and there are few established cultural or legal frameworks in place to deal with such issues. Privacy is especially important in the context of massive data because of the potential for finding associations across sets of data. Given this potential, should it be acceptable for a cell-phone company to make available tracking data for a large number of customers for academic research without restrictions or controls? Would it be acceptable for law-enforcement purposes? For map-makers or the news media? What if the phone data were correlated with other information to give a picture of the owners' patterns of activities?

The companion to data privacy is data ownership. If Google were to go out of business, who owns all the stored email data? If Google's email service were sold off as a separate business to an overseas entity, who owns that data, and what country's laws apply? If InstaBook were to decide to sell all of the user-posted pictures in its system and also declare copyright ownership of them, is that acceptable, and do the people who posted the data have any recourse? If a government pays for plot maps of all properties in its jurisdiction, are these maps public or private? Can mapping companies use them for free? Can they be kept from taxpayers of the jurisdiction? Many transit agencies now track their buses in real time. Does the public own that data? Can services access it for free to show arrival times for future buses and other useful information?

Such thorny issues of privacy and ownership will need to be resolved as society continues to collect data on individuals and their activities. It is easy to see why such topics merit a full study of their own by a committee with a broad set of expertise.

While these issues will not be addressed in this report, the committee does hope to see them addressed in complementary studies. Two comments may help to connect this report to future reports that focus on privacy and other issues of public policy. First, the committee believes that it is impossible to achieve absolute levels of privacy while exploiting the data that arise from human activity. There will necessarily have to be a trade-off, one which is based on an assessment of the relative value of privacy when compared with the possible gains from data analysis. For society to agree on the terms of this trade-off, it will be necessary to understand exactly what are the possible gains from data analysis. The possible gains are part of the focus of this report. Second, the focus of this report on computation and inference aims to understand not only what can be achieved from data analysis, but also what cannot be achieved (cf. the earlier discussion of statistical errors above). In the context of privacy considerations, it may be desirable that certain inferences cannot be obtained reliably; thus, a clear understanding of computation and inference will help feed an understanding of mechanisms for achieving privacy.

WHAT HAS CHANGED IN RECENT YEARS?

In 1995 the National Research Council's Committee on Applied and Theoretical Statistics held a workshop to examine the challenges and promises of the ability to process massive data. The workshop was documented in *Massive Data Sets: Proceedings of a Workshop* (NRC, 1996). Comparing the situation depicted in that report with the current situation allows three areas to be highlighted where changes have been particularly noteworthy.

First, there has been a qualitative leap in the amount of data regarding human interests and activities, much of it generated voluntarily via human participation in social media. Crowdsourcing is also a new phenomenon, as are massive multiplayer online games. With the rise of such human-oriented data sources comes a number of technical challenges. For example, social data are often relational, taking the form of networks that link people and objects along a variety of dimensions. Moreover, such data are often fragmentary and subject to a variety of sampling biases. There are also issues of willful misrepresentation and governmental restrictions on access. Finally, human-oriented data often involve natural language and other representations, with a rich underlying semantics, and the inferential problems of interest often involve reasoning about underlying causes and human intentions.

Second, distributed computing systems have become a reality, with major implications for the collection and processing of massive data. The 1995 workshop clearly recognized that existing algorithms for data analysis would not scale to the kinds of data set sizes that were beginning to accrue, even accounting for the ongoing increase in central processing unit speed, and solutions were sought in parallel and distributed processing systems. Such systems have begun to emerge, driven by a variety of technological and economic factors, and have opened up new vistas and new challenges. In particular, cloud computing now allows access to very large computing infrastructures through a network on an as-needed basis. This has led to new trade-offs involving storage, networking, and processing. It is also important to emphasize that the issue is not solely that of distributed computing, but also that of distributed data. Mobile platforms have proliferated, and data often originate on small devices that have bandwidth limitations that limit or preclude data movement. Moreover, the inferential goals for data analysis often involve bringing together multiple data sources that may have been collected independently. For example, as already noted, social media often provide partial and fragmentary perspectives on individuals, and many questions of interest can only be answered if these perspectives are brought into register. In general, many new challenges have arisen involving computational frameworks that are capable of integrating data across spatial, representational, and administrative domains.

Third, many issues involving the geo-temporal nature of data have come to the forefront. For example, a significant fraction of the data on the Internet is in the form of video streams, a trend that is accelerating. Moreover, a growing number of social media and mobile technologies are generating geo- and time-tagged data. Computer networks generate massive data streams. Scientific data often take the form of time series. In such cases, even if an individual time frame does not involve a massive data set, the temporal sequence can quickly overwhelm storage and computing resources. Indeed, it is common in such cases to develop streaming algorithms that attempt to process the data on the fly, avoiding storage. However, the inferential goals associated with such data often involve the discovery and indexing of temporally extended behaviors, and this generally requires some form of storage. It is also the case that many instances of streaming data require real-time or near-real-time processing; examples include the online auctions run for ad placement in search engines and early alert systems for disease outbreaks. This requirement creates new algorithmic challenges where answer *quality* need to be traded off against answer *timeliness*. Finally, many data sets are also indexed by spatial coordinates (an issue emphasized in the 1996 NRC report). This creates new algorithmic challenges where answer quality and timeliness need to be traded off against the geographic granularity of the answer. The overall issue is often that of coping with massive spatio-temporal and geo-temporal data.

Another way to contrast the situation in 1995 with the current situation is to compare the areas of science and technology that were thought to be impacted by massive data issues. Table 1.1 provides a partial listing of these areas, focusing on scientific and engineering fields. Many of the differences depicted in this table can be attributed to the rapid growth in social media, mobile devices, and sensor networks during the past decade and a half.

TABLE 1.1 Scientific and Engineering Fields Impacted by Massive Data

Area Affected in 1995	Area Affected in 2012	Noteworthy Use Cases
Physical sciences	Physical sciences	Astronomy, particle physics
Climatology	Climatology	
Signal processing	Signal processing	
Medicine	Medicine	Imaging, medical records
Artificial intelligence	Artificial intelligence	Natural language processing, computer vision
Marketing	Marketing	Internet advertising, corporate loyalty programs
N/A	Political science	Agent-based modeling of regime change
N/A	Forensics	Fraud detection, drug/human/ CBRNe trafficking
N/A	Cultural studies	Human terrain assessment, land use, cultural geography
N/A	Sociology	Comparative sociology, social networks, demography, belief and information diffusion
N/A	Biology	Genomics, proteomics, ecology
N/A	Neuroscience	fMRI, multi-electrode recordings
N/A	Psychology	Social psychology

NOTE: CBRNe, chemical, biological, radiological, nuclear, enhanced improvised explosive devices; fMRI, functional magnetic resonance imaging; N/A, not applicable.

ORGANIZATION OF THIS REPORT

The statement of task for the study that led to this report reads as follows:

The study will carry out the following tasks:

- Assess the current state of data analysis for mining of massive sets and streams of data,
- Identify gaps in current practice and theory, and
- Propose a research agenda to fill those gaps.

A primary audience for this report is the community of researchers who need to be adept at analyzing massive data. Because, as will be seen, this is an inherently multidisciplinary subject, the report assumes the reader has (or is willing to develop) an understanding of topics in computer science (including databases and distributed systems), statistics, and optimization. Another important audience consists of the research organizations, especially federal funding agencies, which are building capabilities for the analysis of massive data. The report's identification of research challenges should help those organizations target their programs.

Chapter 2 provides an overview of some of ways in which massive data are currently arising in various scientific and technological fields. Focusing on systems and computer architecture issues, it discusses general trends and then turns to several examples: Earth and planetary science, astronomy, biological and medical research, large numerical simulations, telecommunications and networking, social network analysis, and national security. Chapter 3 pursues the systems perspective further, discussing recent developments in parallel and distributed systems, databases, and streaming architectures.

Chapter 4 addresses issues surrounding the temporal nature of data, serving to highlight the fact that many massive data sets arise as temporal streams and that many interesting inferential questions revolve around the detection of temporal trends, changes, and patterns. Moreover, it is often the case that real-time responses are needed.

In Chapter 5, a more general discussion of data representation is provided, including some of the ways in which massive data arrive in raw form and the transformations that are often applied to data, particularly transformations that attempt to reduce the representational complexity of the data.

Chapter 6 turns to a formal treatment of some of the computational complexity issues that arise in the setting of massive data analysis. The discussion focuses on computational resources and the theoretical characterization of trade-offs among these resources.

Chapters 7 and 8 focus on inferential issues. Chapter 7 addresses statistical model-building in the massive data setting, discussing several of the stages in the inferential pipeline, including data cleansing and validation. In Chapter 8 sampling is discussed, focusing on the data-gathering process but also making links to Chapter 5, where sampling is a key methodology for data reduction.

Chapter 9 treats some of the issues that arise when humans are included in the data-analysis loop. This includes crowdsourcing, where humans are used as a source of training data for learning algorithms, as well as visualization, which not only helps humans to understand the output of an analysis, but also allows human input into model revision.

Chapter 10 attempts to bring several of the strands of the report together into a proposal for a taxonomy of some of the major algorithmic problems arising in massive data analysis. The committee hopes that the ideas in this section will serve to organize the research landscape and also provide a point of departure for the design of "middleware" that links high-level inferential goals to the algorithms and hardware needed to achieve those goals.

In accordance with the study's statement of task, Chapters 2 through 10 identify gaps in current theory and practice, and Chapters 3 through 10 propose a number of elements of a research agenda. Finally, Chapter 11 presents the committee's primary conclusions.

REFERENCES

Halevy, A., P. Norvig, and F. Pereira. 2009. The unreasonable effectiveness of data. *IEEE Intelligent Systems* 2:8-12.

NRC (National Research Council). 1996. *Massive Data Sets: Proceedings of a Workshop.* National Academy Press, Washington, D.C.

2

Massive Data in Science, Technology, Commerce, National Defense, Telecommunications, and Other Endeavors

WHERE ARE MASSIVE DATA APPEARING?

Experiments, observations, and numerical simulations in many areas of science nowadays generate terabytes of data and, in some cases, are on the verge of generating many petabytes. This rapid growth heralds an era of "data-centric science," which requires new paradigms addressing how data are captured, processed, discovered, exchanged, distributed, and analyzed. While traditional methods of analysis have largely focused on analysts being able to develop and analyze data within the confines of their own computing environment, the growth of big data is changing that paradigm for many disciplines, especially in cases in which massive amounts of data are distributed across locations. The distributed and heterogeneous nature of the data provides substantial challenges for many disciplines in the physical and life sciences and also in commerce, medicine, defense, finance, telecommunications, and other industries.

The fact that scientific data sets across a wide range of fields are multiplying is an important driver for modern science. Analyses of the information contained in these data sets have already led to major breakthroughs in fields ranging from genomics to astronomy and high-energy physics, encompassing every scale of the physical world. Yet much more remains, and the great increase in scale of the data creates complex challenges for traditional analysis techniques.

It is not only experimental measurements that are growing at a rapid pace. As stated in Szalay (2011, p. 34): "The volume of data produced by computer simulations (used in virtually all scientific and engineering disciplines today) is also increasing at an even faster rate. Intermediate simulation steps must often be preserved for future reuse because they represent substantial computational investments. The sheer volume of these data sets is only one of the challenges that scientists must confront." Data analyses in some other disciplines (e.g., environmental sciences, wet laboratories in life sciences) are challenged to work for thousands of distinct, complex data sets with incompatible formats and inconsistent metadata.

While the scientific community and the defense industry have long been leaders in generating large data sets, the emergence of e-commerce and massive search engines has led other sectors to confront the challenges of massive data. For example, Google, Yahoo!, Microsoft, and other Internet-based companies have data that are measured in exabytes (10^{18} bytes). The availability and accessibility of these massive data sets is transforming society and the way we think about information storage and retrieval.

Social media (e.g., Facebook, YouTube, Twitter) have exploded beyond anyone's wildest imagination, and today some of these companies have hundreds of millions of users. Social-media-generated texts, images, photos, and videos comprise an unexpected and rapidly growing corpus of data. Data mining of these massive data sets is transforming the way we think about crisis response, marketing, entertainment, cybersecurity, and national intelligence. New algorithms that assess these data in ways other than counting hits on key words, such as the analysis of social relationships, involves large graph analyses and requires new scalable algorithms.

Understanding and characterizing typical Web behavior dynamically (because the time scale of changes on the Internet is in minutes) presents remarkable challenges. In this cyber-oriented world, behavior that does not fit the patterns is often related to malware or denial-of-service attacks. Recognizing these in time, estimating the impact on human behavior, and responding is a new and emerging challenge that has few parallels in science.

Capturing and indexing the Internet has created whole sets of new industries. Some of the world's largest companies are trading in information and have built their business model on appropriately customized advertisements. Interpreting user behavior and providing just-in-time advertisements customized to the users' profiles require very sophisticated data management capabilities and efficient algorithms. Service-sector companies specializing in Internet-based auctions, like eBay or Amazon, have developed sophisticated analytics capabilities. Almost all Web-based companies today are capturing user actions, even if they do not immediately analyze them. This confluence of technologies has created a whole new industry, one based inherently on massive data.

CHALLENGES TO THE ANALYSIS OF MASSIVE DATA

A number of challenges exist in both data management and data analysis that require new approaches to support the "big data" era. These challenges span generation of the data, preparation for analysis, and policy-related challenges in its sharing and use. Initiatives in research and development that are leading to improved capabilities include the following:

- Dealing with highly distributed data sources,
- Tracking data provenance, from data generation through data preparation,
- Validating data,
- Coping with sampling biases and heterogeneity,
- Working with different data formats and structures,
- Developing algorithms that exploit parallel and distributed architectures,
- Ensuring data integrity,
- Ensuring data security,
- Enabling data discovery and integration,
- Enabling data sharing,
- Developing methods for visualizing massive data, and
- Developing scalable and incremental algorithms.

As data volumes increase, the ability to perform analysis on the data is constrained by the increasingly distributed nature of modern data sets. Highly distributed data sources present challenges due to diverse natures of the technical infrastructures, creating challenges in data access, integration, and sharing. The distributed nature also creates additional challenges due to the limitations in moving massive data through channels with limited bandwidth. In addition, data produced by different sources are often defined using different representation methods and structural specifications. Bringing such data together becomes a challenge because the data are not properly prepared for data integration and fusion, and the technical infrastructures lack the appropriate information infrastructure services to support analysis of the data if it remains distributed. Statistical inference procedures often require some form of aggregation that can be expensive in distributed architectures, and a major challenge involves finding cheaper approximations for such procedures. Finally, security and policy issues also limit the ability to share data. Yet, the ever-increasing generation of data from medicine, physical science, defense, and other industries require that analysis be performed on data that are captured and managed across distributed databases.

In addition to challenges posed by the distributed nature of most massive data, the increase of data can also limit, in other ways, the amount of analysis that can be performed. For example, some data require high-performance computational infrastructures for data preparation before analysis can even begin, and access to such capabilities may be limited. An example would be an Earth science investigation that requires first converting the data to a common spatial grid. In cases where the analysis depends on such an expensive pre-processing step, the usefulness of the massive data is enhanced if the data-collection system is engineered to collect data of high quality in forms that are ready for analysis without the pre-processing.

However, rather than having data pre-processed for all scenarios, and thus taking up substantial storage, ad hoc investigations may require that data be processed, and thus sufficient computing infrastructures must be in place to support such ad hoc analysis. This can be desirable scientifically, because understanding of what data are needed may become clearer as an investigation proceeds. In order to support this evolutionary cycle, software systems that handle massive data must be inherently information-driven. That means that their content should be based on explicit information models that capture rich semantics that improve the provenance and understanding of the data. This in turn makes it easier to correlate distributed data sets, thus improving the ability to effectively search large collections of data without requiring changes and updates to the software (and hardware) as the data model evolves and changes during the scientific process.

Finally, challenges exist in better visualizing massive data sets. While there have been advances in visualizing data through various approaches, most notably geographic information system-based capabilities, better methods are required to analyze massive data, particularly data sets that are heterogeneous in nature and may exhibit critical differences in information that are difficult to summarize. This topic is discussed in Chapter 9.

TRENDS IN MASSIVE DATA ANALYSIS[1]

While improvements in computer hardware have enabled today's explosion in data, the performance of different architectural components increases at different rates. Central processing unit (CPU) performance has been doubling every 18 months, following Moore's Law. The capacity of disk drives is doubling at a similar rate, somewhat slower than the original Kryder's Law prediction (Walter, 2005), driven by higher density platters. On the other hand, the disks' rotational speed has changed little over the past 10 years. The result of this divergence is that while sequential input/output (I/O) speeds slowly increase with density, random I/O speeds have changed only moderately. Because of the increasing difference between the sequential and random I/O speeds of disks, only sequential disk access is

---

[1] The first four paragraphs of this section follow Szalay (2011).

possible—if a 100-terabyte (TB) computational problem requires mostly random access patterns, it cannot be done. Finally, network speeds, even in the data center, are unable to keep up with the increases in the amount of data. Said differently, with petabytes (PB) of data, we cannot move the data to where the computing is; instead, we must bring the computing to the data. More discussion of hardware and software for managing massive data is found in Chapter 3.

The typical analysis pipeline of a data-intensive scientific problem starts with a low-level data access pattern during which outliers are filtered out, aggregates are collected, or a subset of the data is selected based on custom criteria. The more CPU-intensive parts of the analysis happen during subsequent passes. Such analyses are currently often implemented in research environments in small clusters of linked commodity computers (e.g., a "Beowulf cluster") that combine compute-intensive, but storage- and I/O-poor, servers with network-attached storage. These clusters can handle problems of a few tens of terabytes, but they do not scale above 100 TB because they are constrained by the very high costs of petabyte-scale enterprise storage systems. Furthermore, as these traditional systems grow to meet modern data analysis needs, we are hitting a point where the power and space requirements for these systems exceed what is available to individual investigators and small research groups.

Existing supercomputers are not well suited for data-intensive computations either, because while they maximize CPU cycles, they lack I/O bandwidth to the mass storage layer. Moreover, most supercomputers lack disk space adequate to store petabyte-size data sets over the multi-month periods that are required for a detailed exploratory analysis. Finally, commercial cloud computing platforms are not the answer either, at least not today. The data movement and access fees are excessive compared to purchasing physical disks, the I/O performance they offer is substantially lower (e.g., 20 megabytes per second), and the amount of provided disk space (often in the range of, say, 10-50 gigabytes) is woefully inadequate for massive data.

Based on these observations, it appears that there is an unmet need today for capabilities to enable data-intensive scientific computations: an inexpensive yet efficient product for data-intensive computing in academic environments that is based on commodity components. The current situation is not scalable and not maintainable in the long run. This situation is analogous to the one that led to the development of the Beowulf cluster.

As data sets are growing at, or faster than, Moore's Law, they are growing at least as fast as computing power increases. This trend tends to limit analytical techniques to those that scale, at most, linearly with the number of data points (N), although those that scale as $N \log N$ are also acceptable because the log N factor can be made up through parallelism. It becomes increasingly difficult to tackle computationally challenging data analyses using existing algorithmic tools, and there is a need to develop new tools that target near-linear computational complexity.

The large-data analytics and e-commerce companies have spent substantial resources on creating a hardware/software framework with performance that scales well with massive data. The typical approach is to create large data centers consisting of hundreds of thousands of low-end computers managed with an extreme economy of scale. Their main features include the following:

- Centralization of large-scale infrastructure,
- Data sizes at the petabyte to exabyte scale,
- Architectures that are highly fault-tolerant, and
- Computing that is collocated with the data for large-scale analytics.

While large businesses in the past have used relational databases, these do not scale well to such extreme sizes. Industries dealing with big data are reacting to data that is more distributed, heterogeneous, and generated from a variety of sources. This is leading to new approaches for data analysis and the demand for new computing approaches. Various innovative data management solutions have emerged, many of which are discussed in Chapter 3. These models work well in the commercial setting, where enormous resources are spent on harvesting and collecting the data through actions such as Internet crawling, aerial photos for geospatial information systems, or collecting user data in search engines. Some of the technical trends that have been occurring to address the data challenges include the following:

- Distributed systems (access, federation, linking, etc.),
- Technologies (MapReduce algorithms, cloud computing, Workflow, etc.),
- Scalable infrastructures for data- and compute-intensive applications,
- Service-oriented architectures,
- Ontologies, models for information representation,
- Scalable database systems with different underlying models (relation to triple stores),
- Federated data security mechanisms, and
- Technologies for moving large data sets.

Many of these technologies are being used to drive toward more systematic approaches.

As discussed in the examples below, many groups are setting up tools that support pipeline or workflow approaches to data analysis. Rather than constructing one large database, the general concept is to enable analysis by bringing together a variety of tools that allow for capture, preparation, management, access, and distribution of data. This collection of tools is configured as a series of steps that constitute a complex workflow for generating and distributing data sets. Such a pipeline may also extract data from operational databases and systems and put that data into environments where it can be prepared and fused with other data sets and staged into systems that support analysis. Challenges include co-utilization of services, workflow discovery, workflow sharing, and maintaining information on metadata, information pedigree, and information assurances as data moves through the workflow.

Several multi-organizational groups are addressing the big-data issue and helping to bring the required scientific expertise and attention to this important problem. Examples are the groups that organize the Extremely Large Databases workshop series,[2] those working in the database research community, and those that have produced open-source technologies such as Hadoop (see Chapter 3). As the state of research in data-intensive systems is focusing on new ways to process data, these groups are leveraging open-source technologies and similar approaches to handle the orchestration of data-processing algorithms and the management of massive amounts of heterogeneous data. A major challenge in the big-data area is in evaluating ways in which distributed data can be analyzed and in which scientific discoveries can be made. Many of the aforementioned technologies, such as Hadoop, and also the proposed SciDB database infrastructure for science data management, have been co-developed across several organizations (both research laboratories and industry) and deployed across organizational boundaries to support analysis of massive data sets. The challenge is both one of developing appropriate systems as well as creating novel methods to support analysis, particularly across highly distributed environments.

In many ways, the big-data problem is simplified when data are centrally stored. However, modern data sets often remain distributed because of technical, social, political, and economic reasons, increasing the challenge to efficiently analyze the data and driving the need to build virtual data systems that integrate assets from multiple organizations. In this emerging paradigm, these virtual systems require elasticity, separation of concerns, scalability, distribution, and information-driven approaches.

Cloud computing is offering an attractive means to acquire computational and data services on an "as needed" basis, which addresses the need for elasticity. This option requires, however, that systems are architected to take advantage of such an infrastructure, which is not often the case. Many systems must be rethought from first principles in order to better exploit the possibilities of cloud computing. For example, systems that decouple data storage, data management, and data processing are more likely to rap-

---

[2] See the Extremely Large Databases website at http://www.xldb.org.

idly take advantage of elastic technology paradigms like cloud computing because they are more suited to leveraging the generic services that cloud computing can readily provide.

EXAMPLES

Earth and Planetary Observations

In the physical sciences domain, there is an increasing demand for improving the throughput of data generation, for providing access to the data, and for moving systems toward greater distribution, particularly across organizational boundaries. Earth, planetary, and astrophysics missions, for example, have all seen an order-of-magnitude increase in data over the past decade, rising from hundreds of gigabytes in some cases to well into the tens to hundreds of terabytes range. The Square Kilometer Array project, as an example, is predicted to produce hundreds of terabytes a second![3] Studies released by each of these disciplines in their decadal reports suggest that this trend will continue. Many Earth and planetary missions have instruments that will continue to generate observations of massive size that will substantially increase the scientific archives worldwide. Climate research also continues to grow at a rapid pace as climate models and satellite observations grow and are needed to support new discovery. NASA's Earth Science enterprise, for example, now manages data collections in the several-petabyte range. In planetary science, the amount of data returned from robotic exploration of the solar system between 2002 and 2012 was 100 times the amount of data collected from the previous four decades. Astrophysics has seen similar increases, and all of these disciplines have experienced a continued increase in the geographic distribution of the data.

As already noted, the increase of data within these science environments has, in many cases, led to the construction of data "pipelines" that acquire data from instruments, process and prepare that data for scientific use, capture the data into well-engineered data management systems, and then provide ad hoc services for data distribution and analysis. Such infrastructures have required advanced computing capabilities that often span multiple institutions and support well-orchestrated information services.

While, traditionally, many of these workflows are constructed for each instrument or mission, there is an increasing interest in performing analysis across data sets that may span different instruments or missions, even from

---

[3] This unprecedented increase will require innovation beyond our current understanding of computation and storage over the next 10 years to achieve the project's ambitious requirements and science goals. Background on the Square Kilometer Array project may be found at http://www.skatelescope.org.

different disciplines. This type of data integration and inter-comparison is not limited to just observational data sets. Within the climate research community, effort is under way to prepare observational data so that it can be inter-compared against the output of climate models. However, within the climate discipline, as with others, data from different institutions, systems, and structures use different standards and measurements, and this lack of standardization makes such analysis difficult, due to the heterogeneity.

Astronomy[4]

Astronomy is a good example for studying how the data explosion happened and how long it might continue. In this realm, successive generations of exponentially more-capable sensors, at the same cost, are the reason for the data explosion (all being traceable to advances in semiconductor technology and, ultimately, to Moore's Law). New generations of inexpensive digital cameras come out every 6 months, and satellites have ever-higher resolution and more pixels. Even old telescopes are getting new instruments that collect more data. For example, the Dark Energy Survey is building a huge array of charge-coupled devices (CCDs) to be placed on an older telescope in Chile.

However, not every domain of science has such growth areas. One could argue that optical astronomy will soon reach the point when increasing the size of collector arrays will become impractical, and the atmospheric resolution will constrain the reasonable pixel size, causing a slow-down in data collection. But time-domain astronomy is emerging, and by taking images every 15 seconds, even a single telescope (e.g., the Large Synoptic Survey Telescope) can easily generate data that can reach 100 PB in a decade.

New instruments and new communities emerge to add to the big-data movement. Radio astronomy, with focal plane arrays on the horizon, is likely to undergo a paradigm shift in data collection, resembling the time when CCDs replaced photographic plates in optical telescopes. Amateur astronomers already have quite large, cooled CCD cameras. When a community of 100,000 people starts collecting high-resolution images, the aggregate data from amateur astronomers may easily outgrow the professional astronomy community.

Biological and Medical Research

A substantial amount of analysis is being performed using data collected by medical information systems, most notably patient electronic health records. This information represents a wealth of data both to im-

---

[4] This section is adapted from Szalay (2011).

prove individual health-care decisions as well as to improve the overall health-care delivery enterprise. The U.S. Food and Drug Administration, for example, is building an active drug safety surveillance system utilizing data from de-identified medical record databases. Medical researchers are gathering together to share information about interventions and outcomes in order to perform retrospective analysis, and insurance companies continue to mine data to improve their own models.

The genomics revolution is proceeding apace, with the cost of sequencing a single human genome soon to drop below $1,000. As the cost decreases, it becomes feasible to sequence multiple genomes per individual, as is being envisaged in tumor genomics initiatives. Overall, data volumes in genomics are growing rapidly. The Short Read Archive at the National Center for Biological Information is soon expected to exceed a petabyte. As more and more high-throughput sequencers are deployed, not just in research but also in hospitals and other medical facilities, we will see an even faster data growth in genomic information.

Neuroscience is increasingly using functional magnetic resonance imaging, where each session can easily result in tens of terabytes of data. Longitudinal studies of hundreds of patients generate data measured in petabytes today. Studies of the cardiovascular systems are creating multiscale simulations from the molecular scale to those of the human circulation system. The European Human Brain Project is setting out to integrate everything we know about the brain in massive databases and detailed computer simulations. And ultra-high-resolution microscopy is also generating very large data sets in cell biology.

Large Numerical Simulations[5]

Numerical simulations are becoming another new way of generating enormous amounts of data. This has not always been the case. Traditionally, these large simulations (such as gravitational N-body simulations in astrophysics or large simulations of computational fluid dynamics) have been analyzed while the simulation was running, because checkpointing and saving the snapshots was overly expensive. This fact traditionally limited the widespread use of simulation data.

Even when a few snapshots have been saved and made public, downloading large files over slow network connections made the analysis highly impractical once simulations reached the terabyte range. The Millennium simulation in astrophysics has changed this paradigm by creating a remotely accessible database with a collaborative environment, following the example of the Sloan Digital Sky Survey SkyServer. The Millennium data-

---

[5] This section is adapted from Szalay (2011).

base drew many hundreds, if not thousands, of astronomers into analyzing simulations as easily as if the observational data were publicly available.

The emerging challenge in this area is scalability. The Millennium has 10 billion particles. The raw data is about 30 TB, but the 1 TB database does not contain the individual dark matter particles, only the halos, subhalos, and the derived galaxies. Newer simulations are soon going to have a trillion particles, where every snapshot is tens of terabytes, so the data problem becomes much worse. At the same time, there is an increasing demand by the public to get access to the best and largest simulations; it is inevitable that the Millennium model is going to proliferate. There may be a need for a virtual observatory of simulations that can provide adequate access and the ability to do analysis, visualization, and computations of these large simulations remotely. This need cuts across all disciplines, because simulations are becoming more commonplace in all areas of physical science, life sciences, economics, and engineering.

As data become increasingly unmovable, the only way to analyze them is "in place." Thus, new mechanisms are needed to interact with these large simulations, because it is not feasible to simply download raw files. For interactive visualizations, it would be easier to send a high-definition, three-dimensional video stream to every interested scientist in the world than it would be to move even a single snapshot of a multi-petabyte simulation from one place to another.

New and interesting paradigms for interacting with large simulations are emerging.

In a project related to isotropic turbulence, data are accessed via a Web service where users can submit a set of about 10,000 particle positions and times and then retrieve the interpolated values of the velocity field at those positions. This can be considered as the equivalent of placing small "virtual sensors" into the simulation instead of downloading all the data or significant subsets of it. The service is public and is typically delivering about 100 million particles per day worldwide. Several papers appearing in the top journals have used this facility.

Telecommunications and Networking

Managing a modern globe-spanning highly reliable communications network requires extensive real-time network monitoring and analysis capabilities. Network monitoring and analysis is used for tasks such as securing the network from intruders and malefactors, rapid-response troubleshooting to network events, and trend prediction for network optimization.

Large telecommunication providers such as AT&T and Verizon offer a wide range of communication services, ranging from consumer (mobile voice and data, wired broadband, television over Internet protocol, and plain old telephone service) to business (data centers, virtual private networks, multiprotocol label switching, content caching, media broadcast), to a global long-haul communications backbone. Each of these services contains many interacting components, and understanding network issues generally requires that data about the interacting components be correlated and analyzed. Further, the global network contains tens of thousands of network elements distributed worldwide.

In addition to the large volumes of data involved, the major problem in telecommunications and networking data analysis is the complexity of the data sets (hundreds to thousands of distinct data feeds) and the requirement for real-time response. Monitoring the health of tens of thousands of backbone routers generates large data sets, but backbone routers are only one component of the global end-to-end communications network. Many user applications—for example, streaming music to a smartphone—require the correct and efficient operation of dozens of different network elements. Troubleshooting a bad connection requires that dozens of data feeds be monitored and correlated.

Responding to network events, such as misbehaving routers, stalled routing convergence algorithms, and so on, or to network intruders, requires fast response; delays result in customer dissatisfaction, or worse. Therefore, all of the hundreds of data feeds must be managed in a real-time warehouse, which can provide timely answers to troubleshooting queries. The technology for such real-time warehousing is still being developed.

Social Network Analysis

Social network analysis is the science of understanding, measuring, and predicting behavior from a relational or structural perspective. Using a blend of graph-theoretic and nonparametric statistical techniques, researchers in this area take data, such as phone data or observations indicative of interpersonal connections, and identify who are the key actors and key groups within and across networks, also identifying special patterns and important pathways. Traditional data sets focused on information within groups—such as who worked with whom or who is friends with whom—or between groups, such as which countries are allied or which organizations supplied goods to which others. The network in those traditional investigations might be a simple one- or two-mode network (i.e., a network with nodes that fall into just one or two classes), and the links were often binary. Often the data were from a small contained group.

Today the field of social network analysis has exploded, and data often take the form of meta-networks in which information about who, what, where, how, why, and when are linked using multi-mode, multi-link, multi-level networks. (See, e.g., Carley, 2002.) The links are probabilistic, and the nodes have states that may change over time. The size of networks of interest are often larger than in the past, such as the entire citation network in the web-of-science or the network of phone calls in all countries over a period of 12 months. The availability of, and interest in, such massive network data has increased as social media sites have become more prevalent; as data records for public functions (such as home sales records and criminal activity reports) have increasingly been made public; and as various corporations make such data available, at least in anonymized form (such as phone records, Internet movie databases, and the web-of-science).

Massiveness for network data arises on several fronts. First, the number of nodes in data of interest has grown from hundreds to several million or billion. Second, the number of classes of nodes that need to be included in a single analysis has grown. For example, web-of-science data have authors, topics, journals, and institutions, with each of these "nodes" having multiple attributes. Third, the data are often collected through time and/or across regions. For example, Twitter, Lexis-Nexis, AIS, and various sensor feeds all have networks embedded in space and time. In particular, social media contain meta-network data that are massive and still growing.

Each source of massiveness presents the following technical challenges:

- The increase in the number of nodes whose data are being analyzed means that many of the traditional algorithms must be replaced by ones that scale better. This has been easy for metrics that rely only on measures associated with individual nodes and their direct links to other nodes—i.e., on local information for each node. For example, degree centrality, the number of links to/from a node, scales well, and it also can be readily adapted to streaming data. However, algorithms that do not scale well are those (such as betweenness) that rely on examining paths through an entire network or on local simulations that use multi-mode, multi-link data. In general, many single-node algorithms that include key node identification and grouping algorithms either scale well, are already parallelized, or have heuristic-based approximation approaches (Pfeffer and Carley, 2012). The main challenges here are rapid re-estimation given streaming data, estimating emergence and degradation of a node's position over time given dynamic data with missing information, comparison of multiple networks, and enumeration of motifs of interest.

- Multi-mode data are challenging in that there are few metrics, and new ones are needed for each application. To be sure, a set of metrics exists for two-mode networks; however, most of the massive data is n-mode. From a massive data perspective, the key challenge is that the search paths tend to increase exponentially with the number of node classes (i.e., modes). Improved metrics, scalable multi-mode clustering algorithms, improved sets of interpretations, and improved scaling of existing metrics are the core challenges.

- For temporal data there are two core challenges: incremental assessment and atrophication/emergence. Incremental assessment requires new algorithms to be defined that reflect social activity, which can be rapidly computed as new data become available. Even newly developed incremental algorithms are still exponential with network size; more importantly, it is not clear that the existing metrics (e.g., betweenness), even if sufficiently scalable incremental algorithms were to be developed, are meaningful in truly massive networks. Other challenges center around the problem of identifying points of atrophication and emergence, where portions of the network are fading away or emerging. The identification of simple temporal trends is not particularly a challenge, as Fourier analysis on standard network metrics provides guidance and scales well. With temporal data such as email and Twitter, not all nodes (the people who send information) are present in every time period. Understanding whether this lack of presence represents missing data due to sampling, temporary absence, or to a node actually leaving the network is a core challenge.

Thus, the harder problems that are particularly impacted by massive data include (1) identifying the leading edge in a network as it is being activated (e.g., who is starting to contract a disease or where is a revolution spreading), (2) identifying what part of the network is contributing to anomalous behavior, and (3) updating metrics as data changes. Geo-temporal network data present still further challenges, due in part to the infrastructure constraints that inhibit transmitting and sharing geo-images and the lack of large-scale, well-validated, spatial data for locations of interest in network analyses. However, even if these technical and data problems were solved, dynamic geo-enabled network analysis would still be problematic due to the lack of a theoretical foundation for understanding emergence in spatially embedded networks. A major problem in this area is diffusion. Well-validated spatial and social network models exist for the spread of distinct entities such as disease, ideas, beliefs, technologies, and goods and services. However, these models often do not make consistent predictions; that is, spatial and social network models disagree, and they typically do not operate at the same scale. Having an integrated spatial-social-network model for the diffusion of each type of entity is critical in many areas. The increase in geo-temporal tagged network data is, for the first time, making it possible to create, test, and validate such models; however, current mathematical and computational formulations of these models do not scale to the size of the current data.

Much current network analysis is done on individual standalone single-processor machines. However, that is changing. There are a few tools (e.g., the Organizational Risk Analyzer (ORA) toolkit) that have parallelized the algorithms and make use of multi-processors. In addition, there are Hadoop versions for some of the basic algorithms,[6] thus enabling utilization of cloud computing. Many diffusion routines can now be run on Condor clusters. In general, there is a movement to distributed computing in this area; and although the trend will continue, the existing technologies for network analysis in this area are in their infancy. New tools are appearing on a regular basis, including special tools for supercomputers, tools that take advantage of special processors, chips with built-in network calculations, and algorithms that utilize the memory and processors in the graphics display.

Social networking—the use of a social media site to build, maintain, review, and disseminate information through connections—is a growing trend. Such sites include Twitter, Facebook, LinkedIn, and YouTube, and they are a growing source of massive data. In general, social media are being applied as a means to gather, generate, and communicate data in a rapid fashion. They are used in marketing by companies to announce products and collect consumer feedback and also by companies to discern other companies' secrets. They are used by groups to organize social movements, protests, track illegal activity, record the need for social services (e.g., pothole filling and snow removal), provide feedback on quality of restaurants, hotels, services, and safety of areas, and so on. Most technologies applied to such social media do little more than collect data—sometimes providing simple visualizations—and simply count the frequency of messages, key words, hashtags, etc. Truly making use of these data requires scalable clustering techniques, real-time ontology abstraction, and on-the fly thesauri creation for extracting the complete network associated with a topic of interest.

Social network analysis technologies can be used to assess social media data, while social network theory can be used to address how people will connect via social media and how it will change the nature of their interactions. However, many challenges remain. Social network analysis has traditionally focused on small, complete networks (i.e., fewer than 100 members, in which all members were contacted), where interaction can be face-to-face, and all data come from one time period. To exploit social media data, techniques have to be expanded to handle large networks (e.g., thousands or millions of nodes), where the data are sampled rather than

---

[6] See, for example, X-Rime, available at http://xrime.sourceforge.net.

complete (so there may be sampling biases in what data are captured), and for which the data are typically dynamic and the dynamics are of interest. Hence, issues of missing data, link inference, bias, forecasting, and dynamics are now of great interest. Also, network metrics are highly sensitive to missing and erroneous data, and so alternative metrics, confidence intervals on existing metrics, and procedures for inferring missing data are all needed. Advances in these areas are occurring rapidly. An emerging challenge, however, is how one can cross-identify the same person in multiple social media. This is particularly important for tracking criminal activity, terrorists, or pedophiles.

Social media are still evolving. As they mature, the shape of the technology itself will be different, and new users that will have grown up with them will be the dominant group. As such, cultural norms of usage will emerge. Changes in security and privacy options on social media sites are liable to make such cross-identification even more challenging than it is currently. Because the technology is a major source for collecting and processing massive data, the needs and challenges facing data analysis are likely to change as the technology matures.

Some of the common questions of analysis with respect to social networks include the following:

- *Effective marketing with social media.* How can companies and governments measure the effectiveness of their social media campaigns, assess change in the resulting culture, and understand when the message changes what people do?[7] What new scalable social network tools, techniques, and measures are needed for identifying (1) key actors for spreading messages, (2) early adopters, (3) the rate of spread, and (4) the effectiveness of the spread, given the nature of social media data, to track crises and identify covert activity?[8]

- *Sentiment assessment monitoring and control.* How can one measure, assess, forecast, and alter social sentiment using diverse social media? What new scalable social-network techniques are needed for assessing sentiment, identifying sources of sentiment, tracking changes in groups and sentiment simultaneously, determining whether the opinion leaders across groups are the same or different, and so on? (Pfeffer et al., Forthcoming).

- *Social change in images.* A vast amount of the data in social media sites is visual—videos and photographs. How can these data be

---

[7] See, e.g., Minelli et al. (2013).

[8] See, e.g., De Choudhury et al. (2010), Wakita and Tsurumi (2007), and Morstatter et al. (2013).

assessed, measured, and monitored in a scalable fashion so as to allow social change to be tracked and individuals identified through a fusion of verbal and visual data? (Cha et al., 2007).

- *Geo-temporal network analysis.* An illustrative problem is the assessment, by area within a city and time period, of the Twitter network during disasters in order to rapidly identify needs and capabilities. Currently there are few geo-network metrics; however, the increasing prevalence of geo-tagged data is creating the need for new metrics that scale well and support drill-down analysis on three dimensions at once—space, time, and size of group. Current clustering algorithms scale well on a single dimension, but scalable algorithms are needed that cluster in space and time. Finally, most spatial visualization techniques take the form of heat maps, which are often too crude to visually convey nuances in the data. But if the fullness of big data is exploited, then the overlay of points on a map becomes confusing, so new visual analytic techniques are needed (Joseph et al., 2012).

The sheer size and complexity of data about social networks, cultural geography, and social media are such that systems need to be designed to meet three goals: automation, ease of use, and robustness. Analysts simply do not have the time to capture and run even basic analyses in a time-sensitive fashion. Hence, the data-collection and analysis processes need to be automated so that information extraction can operate independently, and basic statistics identifying key actors and groups can be continually updated. Transparency must be maintained, and the analysts must be able to check sources for any node or link in an extracted network. An increasing number of jobs require tracking information using social network data, and an increasing number of activities that individuals engage in can be discerned from information on the individual's social network, particularly when multiple networks, multiple types of nodes, and multiple relations can be overlain on one another.

National Security

The rise of the Internet has enormously increased the volumes of data potentially relevant to counterterrorism, counter-proliferation, network security, and other problems of national security. Many problems in national defense involve flows of heterogeneous, largely unstructured data arriving too rapidly to be aggregated in their entirety for off-line analysis.

As an example, consider the problem of computer network defense, or cyberdefense. Network-based attacks on computer systems pose threats of espionage and sabotage of critical public infrastructure. If one can observe network traffic, one may wish to know when an attack is under way, who is conducting the attack, what are the targets, and how a defense may be mounted. Both real-time and post-hoc (forensic) capabilities are of interest. The metadata associated with the traffic—e.g., entity X communicated with entity Y at time T using protocol P, etc.—can be regarded as a dynamic graph or arrival process whose analysis may be useful.

To cope with such problems, advances are needed across the board, from statistics to computer system architecture, but three areas can be highlighted. First, streaming algorithms that can process data in one pass with limited memory are clearly important. Second, for data at rest, transactional databases are generally not needed, but highly usable systems for hosting and querying massive data, including data distributed across multiple sites, will be essential. The MapReduce framework, discussed in Chapter 3 of this report, is perhaps a good first step. Third, better visualization tools are also needed to conserve the scarce and valuable time of human analysts.

Two areas in national security that are particularly impacted by massive data are the potential capability for the remote detection of weapons of mass destruction and improved methods of cyber command and control. Although many details of national security problems are classified, approaches to these problems often parallel current efforts in academia and industry.

REFERENCES

Carley, K.M. 2002. Smart agents and organizations of the future. Chapter 12, pp. 206-220 in *The Handbook of New Media* (L. Lievrouw and S. Livingstone, eds.). Sage, Thousand Oaks, Calif.

Cha, M., H. Kwak, P. Rodriguez, Y.-Y. Ahn, and S. Moon. 2007. I tube, you tube, everybody tubes: Analyzing the world's largest user generated content video system. Pp. 1-14 in *Proceedings of the 7th ACM SIGCOMM Conference on Internet Measurement* (IMC '07). ACM, New York, N.Y.

De Choudhury, M., Y.-R. Lin, H. Sundaram, K.S. Candan, L. Xie, and A. Kelliher. 2010. How does the data sampling strategy impact the discovery of information diffusion in social media? *Proceedings of the Fourth International AAAI Conference on Weblogs and Social Media*. Available at http://www.aaai.org/ocs/index.php/ICWSM/ICWSM10/paper/viewFile/1521/1832.

Joseph, K., C.H. Tan, and K.M. Carley. 2012. Beyond "local," "social," and "category": Clustering Foursquare users using latent "topics." In *4th International Workshop on Location-Based Social Networks* (LBSN 2012) at UBICOM, September 8, 2012, Pittsburgh, Pa.

Minelli, M., M. Chambers, and A. Dhiraj. 2013. *Big Data, Big Analytics: Emerging Business Intelligence and Analytic Trends*. John Wiley and Sons, Hoboken, N.J.

Morstatter, F., J. Pfeffer, H. Liu and K.M. Carley. 2013. Is the sample good enough? Com-
    paring data from Twitter's streaming API with Twitter's firehose. In *Proceedings of
    International AAAI Conference on Weblogs and Social Media* (ICWSM), July 8-10,
    Boston, Mass.
Pfeffer, J., and K.M. Carley. 2012. k-centralities: Local approximations of global measures
    based on shortest paths. Pp. 1043-1050 in *Proceedings of the WWW Conference 2012*.
    First International Workshop on Large Scale Network Analysis (LSNA 2012), Lyon,
    France. Available at http://www2012.wwwconference.org/proceedings/index.php.
Pfeffer, J., T. Zorbach, and K.M. Carley. 2013, Forthcoming. Understanding online firestorms:
    Negative word of mouth dynamics in social media networks. *Journal of Marketing
    Communications*.
Qualman, E. 2013. *Socialnomics: How Social Media Transforms the Way We Live and Do
    Business*. John Wiley and Sons, Hoboken, N.J.
Szalay, A.S. 2011. Extreme data-intensive scientific computing. *Computing in Science and
    Engineering* 13(6):34-41.
Wakita, K., and T. Tsurumi. 2007. Finding community structure in mega-scale social networks.
    Pp. 1275-1276 in *Proceedings of the 16th International Conference on World Wide Web*
    (WWW '07). ACM, New York, N.Y.
Walter, C. 2005. Kryder's Law. *Scientific American*, August. pp. 32-33.

3

Scaling the Infrastructure for Data Management

Very-large-scale data sets introduce many data management challenges. Two notable ones are managing the complexity of the data and harnessing the computational power required to ingest and analyze the data. Tools for analyzing massive data must be developed with an understanding of the developing capabilities for management of massive data.

SCALING THE NUMBER OF DATA SETS

Increasing the number of data sets brought to bear on a given problem increases the ability to address the problem, at least in principle. Sets that add rows (e.g., adding more patient records in a health-care data set) to existing data tables can increase their statistical power. Sets that add columns (e.g., adding the patient's smoking history to each patient record) can enable new applications of the data. In the health-care domain, cross-collection data mining can enable exciting advances in personalized medicine. Another example is in the finance domain, where an ability to evaluate loans across multiple banks (e.g., as mortgages are bought and sold) is more powerful than an analysis that is limited to the records of just one bank.

Unfortunately, scaling the number of data sets is very difficult in practice due to heterogeneity in data representations and semantics, data quality, and openness. These aspects are explored below.

Data Representation and Semantics

Data sets managed by different (sub)organizations tend to have disparate representations and semantics. These attributes are described by metadata, which is critical for ensuring that the data can be effectively interpreted. Metadata consist of both structural and discipline-specific metadata. The structural metadata describe the structures of the data and its organization. The discipline-specific metadata describe the characteristics or uniqueness of the data for a particular discipline.

Many disciplines are working toward defining rich semantic models that can help in data searching and understanding. Ideally, data representation standards would permit improvisation; for example, a standard might stipulate a set of structured fields with a free-form key/value map that accommodates unforeseen information. Such an approach can interpolate between the current extremes of restrictive up-front standardization versus free-form chaos. Rich semantics allows tools to be developed that can effectively exploit relationships, thus enabling improved discovery and navigation, and several standards and technologies are emerging. However, current capabilities are still highly dependent on defining well-formed models and structures up-front. It may be important to consider how to evolve data standards over time so that as patterns are recognized in free-form entries, they can be gradually folded into the structured portion of the representation.

An alternative to requiring extensive metadata up-front is to aim for more of a "data co-existence" approach, sometimes referred to as a "dataspace." A good description of that concept is captured in the following Wikipedia entry:

> [Dataspaces provide] base functionality over all data sources, regardless of how integrated they are. For example, a [system] can provide keyword search over all of its data sources, similar to that provided by existing desktop search systems. When more sophisticated operations are required, such as relational-style queries, data mining, or monitoring over certain sources, then additional effort can be applied to more closely integrate those sources in an incremental ["pay-as-you-go"] fashion. Similarly, in terms of traditional database guarantees, initially a dataspace system can only provide weaker guarantees of consistency and durability. As stronger guarantees are desired, more effort can be put into making agreements among the various owners of data sources, and opening up certain interfaces (e.g., for commit protocols).[1]

---

[1] The Dataspaces entry is available at http://en.wikipedia.org/wiki/Dataspaces, accessed May 8, 2013.

In a sense, this approach postpones the labor-intensive aspects of data fusion until they are absolutely needed.

Data Quality and Provenance

Real-world data sets vary in quality, due to a range of factors, including imperfect data collection instruments, human data entry errors, data fusion mistakes, and incorrect inferences. When dealing with a large number of data sets from diverse sources, systematic recording and tracking of data-quality metadata are very important. Unfortunately, in full generality that goal appears to be extremely challenging.

A less daunting but still very ambitious goal is to track the provenance of data elements—that is, their origin, movement, and processing histories. Provenance is also useful for purposes other than reasoning about quality, such as in propagating data updates efficiently, and to attribute data properly in scientific publications. Provenance capture, representation, and querying are active research topics in the communities dealing with scientific workflow, file systems, and databases. The three communities' approaches emphasize different priorities among various trade-offs (e.g., the trade-off between capture overhead and query expressiveness). None of these approaches has reached significant levels of adoption in practice. Currently, provenance is managed with one-off approaches and standards—for example, in planetary science research, each observation is tagged with the time, location, and platform from which it originated—and there is little systematic support for propagating provenance metadata with data wherever it travels.

Representation and propagation of constrained forms of data-quality metadata, such as confidence scores and error bars, is also an active area of research, although to date most work in that area has concentrated on theoretical issues. There could be an opportunity to consider how more formal statistical notions of uncertainty might be incorporated. Overall, there has been little work on scalable systems for the management of uncertain data.

Openness

Non-public data sets require great care when being shared across organizations. To maximize the pool of data that can be shared openly, technologies are needed that fuse open data while protecting proprietary data and preserving anonymity requirements. Data being shared that has been derived from private data (e.g., statistics created by aggregating private data points) is especially problematic, due to data leakage issues (accidental leakage as well as "harvesting" by malicious parties).

SCALING COMPUTING TECHNOLOGY THROUGH DISTRIBUTED AND PARALLEL SYSTEMS

Massive data processing, storage, and analysis will require support from distributed and parallel processing systems. Because the processing speed of microelectronics is not increasing as rapidly as it used to, modern central processing units (CPUs) are instead becoming highly parallel. That is the only way to continue the performance improvements in large-scale processing that are demanded by applications.

In order to enable performance improvements in processing, input/output (I/O) and storage must also become parallelized and distributed. Amdahl's Law—a rule of thumb that has been valid for nearly 50 years—states that a balanced system needs one bit of I/O per CPU cycle, and thus improvements in processing speed must be matched by improvements in I/O. And high I/O performance necessitates a heavy use of local (on-chip) data storage, so that storage is as distributed and parallel as is processing.

The reason that single-threaded computation is still so common is that parallel and distributed systems are difficult to configure and maintain, and parallel and distributed software is difficult to write. The end of the ever-faster CPU era has led to once-exotic technologies becoming commonplace and to new parallel programming and data management systems that are easier to use. This section outlines recent trends in parallel and distributed computing and I/O.

Hardware Parallelism

"Hardware parallelism" is defined here as the presence in a system of many separate computing elements that operate simultaneously. In some cases the elements perform highly specialized tasks and, as a result, can do so very quickly with many elements operating in parallel. A long-standing example of hardware parallelism is integrated circuits for signal processing that can perform Fast Fourier Transforms as a hardware operation. More recent developments are motivated by problems in network management and by the hardware developed to accelerate computer graphics.

Specialized networking equipment, such as very-high-speed network monitoring and firewall gear, commonly makes use of specialized hardware known as field programmable gate arrays (FPGAs) and ternary content addressable memory (TCAM). FPGAs are customizable integrated circuits that can be configured for high performance on special-purpose tasks. A TCAM is akin to a cache memory, but TCAM chips allow the user to specify tie-breaking rules in the case of multiple matches. For example, TCAMs store subnetwork address ranges of interest and the FPGAs perform specialized tasks, such as regular expression matching on the packet content and routing (perhaps to a host for monitoring) based on the results of the TCAM and other analyses. FPGAs have also found use in data warehouse analytics engines, performing filtering and other tasks on data streaming from disk.

Another recent development in hardware parallelism is motivated by the graphics processing units (GPUs) developed to accelerate computer graphics. GPUs are highly parallel processors, originally developed for high-end graphics applications and computer games. Various vendors (NVIDIA, ATI, IBM) have developed such platforms, which are increasingly used for applications requiring very high floating-point performance. (Many of the world's top 500 computers are hybrid machines consisting of a large array of traditional CPUs and GPUs.) A typical GPU card can outperform a CPU by up to an order of magnitude, depending on the application, and the performance of a typical high-end graphics card exceeds a teraflop. Sorting performance is also spectacular, exceeding the rate of 1 billion records per second on some benchmarks. The performance per unit power dissipated is also significantly better with GPUs than with traditional processors, an issue that is becoming increasingly important for the total cost of ownership of high-end computing systems.

The main disadvantage in applying GPUs to large-scale data-intensive problems is the rather limited memory (typically 2-3 gigabytes (GB), up to 6 GB currently) attached to the cards. While data access for the on-board memory is very fast, over 100 GB/s, moving data in and out of the cards can be a bottleneck.

Also, GPU programming is still rather complicated, requiring special environments (e.g., CUDA, OpenCL). Because of the single-instruction/multiple-data nature of the hardware, special attention must be paid to laying out the data to match the configuration of the low-level hardware. The situation is getting better every year, as more and more algorithms are ported to the GPU environment, and increasingly sophisticated debugging environments are emerging. Much of the Linear Algebra PACKage library (LAPACK) has been ported, the Fastest Fourier Transform in the West (FFTW) library for performing discrete Fourier transforms is part of the basic CUDA library, and many graph algorithms have also been successfully ported to GPUs.

The hardware is quickly evolving. Upcoming GPU cards will support more generic memory access, better communication with the host, more flexible task switching, and preemptive multitasking, making them increasingly comparable in programmability to traditional multicore architectures. There are many current efforts to integrate GPUs with databases and stream processing systems.

Multicore CPUs

Over the past decade, conventional server-class CPUs have gained internal parallelism in two ways: by growing the number of cores (independent execution engines) per chip or per package and by increasing the number of operations a core can execute per cycle, largely by means of parallel operations on short vectors of data. Both trends increase the challenge of making good use of the available computational resources in real applications. Tools for automatically parallelizing and vectorizing applications are not currently very effective.

To date, the growth in the number of cores per chip has been somewhat restrained by market forces, in particular the need to retain good performance on non-parallel code. But the industry is well aware that higher peak performance, and higher performance per watt, can be achieved by integrating a much larger number of cores running at lower speeds. Whether massively multicore CPUs will soon play a major role in data analysis is hard to predict, but their eventual arrival now seems all but inevitable (Asanovic et al., 2006).

Flash Memory

The rapid proliferation in flash memory is another very relevant trend. As noted in a recent publication on the topic,

> Traditionally, system designers have been forced to choose between performance and safety when building large-scale storage systems. Flash storage has the potential to dramatically alter this trade-off, providing persistence as well as high throughput and low latency. The advent of commodity flash drives creates new opportunities in the data center, enabling new designs that are impractical on disk or RAM infrastructure (Balakrishnan et al., 2012, p. 1).

To date, flash memory has been used as a fast alternative for disk storage, but it appears to be a promising technology for lowering power requirements while maintaining high reliability and speed for large-scale data systems.

Data Stream Management Systems

Data stream management systems (DSMS) have emerged as a significant research topic over the past decade, with many research systems (Stream, Niagara, Telegraph, Aurora, Cougar) and commercial systems (Streambase, Coral8, Aleri, InfoSphere Streams, Truviso) having been developed. A DSMS runs a collection of standing queries on one or more input streams. The source streams are generally real-time reports of live phenomena. Examples include stock ticker and other financial streams, feeds from sensor networks (e.g., highway monitoring), Web click-streams, video streams, streams of data from scientific experiments (e.g., astronomical or high-energy physics observations), and communications network traffic monitoring. These feeds are processed, correlated, and summarized by the DSMS on a continual basis for immediate action or further analysis.

There are two main methods for writing a query set for a DSMS. The first method uses a highly structured query language, often a variant of SQL (Structured Query Language), such as Contextual Query Language, CQL, which was developed for Stream. Another class of common stream query language incorporates regular expression-matching features to perform complex event detection. These stream languages differ from conventional SQL in that they generally require queries to use windowing constructs to limit the scope of the data used to compute any output record. For example, a stream query might ask, "for each five minute period, report the number of distinct source Internet protocol addresses of packets flowing through this network interface."

The second method for specifying a query set to a DSMS uses a graphical "boxes-and-arrows" approach. Boxes represent data-processing tasks (or data sources), and arrows represent data flow. The programmer selects and customizes the boxes, then connects them with arrows to develop a data processing specification—often through a graphical user interface (e.g., Streambase, Infosphere Streams). The motivation for the boxes-and-arrows method of programming is that many stream analyses are difficult to express in an SQL-like language (e.g., time-series analysis for financial applications, facial recognition in video streams). However, a DSMS query set expressed using a structured query language is generally easier to write and maintain, and it can be more readily optimized.

There is not a rigid boundary between language-based and "boxes and arrows" data stream systems, as one can generally incorporate special-purpose operators into a language-based DSMS, and structured language programming tools have been developed for "boxes-and-arrows" DSMSs (e.g., Streambase, InfoSphere Streams).

If the DSMS is programmed using a declarative query language, the query analyzer will convert the textual queries into a collection of stream operators so that in either case a collection of interconnected stream operators is presented to the query optimizer. A directed graph of stream operators presents special opportunities for the query optimizer because a large and long-running system is presented to optimization. A stream query system presents many opportunities for multi-query optimization, ranging from scan sharing to identifying and merging common execution subtrees, which are not normally available in a database management system (DBMS). The well-structured and explicit nature of the data flow in a data stream system can enable highly effective optimizations for parallel and distributed stream systems.

For example, GS Tool from AT&T Labs Research will analyze its query set to determine an optimal hash partitioning of the packet stream from the network interfaces that it monitors. The output of a high-speed interface, such as 10 Gigabit Ethernet, is normally split into multiple substreams. Very-high-speed links (e.g., the optical transmission rate OC-768) are normally split into eight 10-Gigabit Ethernet streams by specialized networking equipment, at the direction of the query optimizer. The InfoSphere Streams system makes many optimizations to the query graph to optimize parallel and distributed processing: splitting streams to enable parallelism of expensive operators, coalescing query operators into processing elements to minimize data copying, and allocating processing element instances to cluster nodes to maximize parallelism while minimizing data copy and network transmission overhead.

Cluster Batch (Grid) Systems

A cluster of high-performance servers can offer powerful computing resources at a moderate price. One way to take advantage of the computational resources of a cluster of servers is to use grid software. Examples of grid systems are Sun Grid and Load Sharing Facility. A typical grid system allows users to submit collections of jobs for execution, and the jobs are queued by the grid job manager, which schedules them for execution on nodes in the cluster. The job manager performs load balancing among the cluster nodes and shares compute resources among the users. A grid system can use a storage area network to create a local file system for a user's job, or it can use a cluster file system.

A cluster file system is a common solution to the challenge of accessing massive distributed data. Such a system provides location-transparent access to data files to the servers on the cluster. The discussion that follows distinguishes between two types of cluster file systems, those which are POSIX compliant (or nearly so), and those which are not.[2]

A POSIX-compliant cluster file system is attractive to programmers because it provides a traditional interface for data access while requiring minimal reworking of a code base to take advantage of a cluster's resources. POSIX-compliant cluster file systems are often built on top of a storage area network, typically one or more racks of hard drives attached to a high-speed network such as Fibre Channel.

---

[2] POSIX refers to the Portable Operating System Interface standards developed by the Institute of Electrical and Electronics Engineers and International Organization for Standardization to ensure compatibility between software and operating systems.

A high-performance cluster file system such as Sun Microsystems' Quick File System (QFS) separates metadata processing from disk block access. A metadata server manages one or more file systems, maintaining directories and I-nodes (perhaps on separate and specialized storage device such as an array of solid-state drives) and serving file metadata requests to the compute clients on the cluster. The clients access the actual data by direct access to the disks of the storage area network. File reliability and availability is typically provided by using the redundant array of independent disks (RAID) technology.

Although a POSIX-compliant cluster file system is intended to be a transparent replacement for a local file system, the complexities of implementing a distributed file system generally result in some gaps in compliance. These gaps generally occur where complex synchronization would be involved, e.g., file locks and concurrent file access by processes on different servers.

The difficulties of providing POSIX-compliance in a very-large-scale cluster have motivated the development of non-POSIX-compliant file systems, for example the Google file system and the Hadoop distributed file system (HDFS). The discussion below is based on the Google file system (Ghemawat et al., 2003), but HDFS is similar.

The Google file system is designed to support distributed analysis tasks, which primarily make scans of very large files. The underlying assumptions of the Google file system are as follows:

- Files are very large and contain well-defined records.
- Files are usually updated by processes that append well-defined records. The sequential ordering of the appended records is not critical, and many processes may be appending records concurrently.
- Analysis processes generally make large sequential scans of the data files.
- Actual files are stored in the local file systems of servers configured into one or more racks in a data center.

As with QFS and similar systems, the Google file system separates the metadata server from the data servers. The metadata server keeps a hot spare in sync by logging all metadata operations to the hot spare before responding to client requests. A single metadata server might coordinate a file system distributed over thousands of nodes, so minimizing the number of metadata requests is critical. Therefore the file block size is very large—64 megabytes (as compared to 4 kilobytes typical on a local file system).

To minimize the overhead and complexity of metadata-server failure recovery, the Google file system makes only weak guarantees about the correctness of the data written into a file. Duplicate records might be written into the file, and the file might contain garbage areas. The clients that use the Google file system must make provisions for these problems: records should contain consistency information (e.g., checksums), and analysis clients must filter out duplicate records. File availability is ensured using replication; a file block is, by default, replicated to three storage hosts, although critical or frequently accessed files might have a higher degree of replication.

An interesting aspect of the Google file system is that a single server with a hot spare controls thousands of file server nodes. This type of control system can be highly reliable because any single server is unlikely to fail. Instead, provisions are made for recovering from failures among the thousands of file servers—a failure among thousands of nodes is far more likely. Synchronization is achieved using lightweight mechanisms such as logging and the use of file leases. Heavyweight synchronization mechanisms such as Paxos are reserved for the file lock mechanism.

MapReduce

MapReduce is a style of distributed data analysis that was popularized by Google for its internal operations (see Dean and Ghemawat, 2004). Hadoop is an open-source version of MapReduce. MapReduce takes its name from a pair of functional programming constructs, map and reduce. A map invocation applies a function to every element of a list, while a reduce invocation computes an aggregate value from a list.

As used in a MapReduce system, the map phase will organize a collection of compute nodes to divide up a data source (e.g., one or more files) and apply the map'ed function to every record in the file(s). The result is the value of the function on the record, as well as a hash value. In the reduce phase, the map results are reshuffled among the compute nodes, using the result hash to ensure that common records get sent to the same node. The reduce node combines records with the same hash into an aggregate value.

As stated, MapReduce would not seem to provide a powerful programming construct. However, in the context of a large cluster, a MapReduce system provides a couple of critical services:

1. The master server that organizes the MapReduce computation hands out portions of the computation to participating nodes and monitors their progress. If a node fails, or is slow to finish, the master will hand the unit of work to another node.
2. The master server will ideally have a map of data locations (especially if the Google file system or the HDFS is used), node locations, and the network interconnecting them. The master server can at-

tempt to assign processing close to the data (same server or same rack), distribute work evenly among servers, coordinate among concurrent jobs, and so on.

By providing reliability and basic optimizations, MapReduce (or Hadoop) greatly simplifies the task of writing a large-scale analysis on distributed data for many types of analyses.

The abstraction offered by a single MapReduce job is rather constrained and low-level, relative to the needs of applications that are moving to MapReduce-based platforms. Such applications range from Web data management to genomics to journalism. As a result, there are numerous efforts to layer more flexible and high-level abstractions on top of MapReduce. Examples include machine-learning libraries (e.g., Mahout), structured query languages (e.g., Jaql, Hive, Pig Latin), and workflow managers (e.g., Cascading, Oozie).

For some application scenarios, using a MapReduce job as a building block is not considered a good fit in terms of system performance considerations such as latency and throughput. Hence, several projects are creating variations on MapReduce. One group of projects offers a general directed-acyclic-graph (DAG) processing model (e.g., Dryad, Hyracks, Nephele). Another group caters to applications that require iterative processing of a data set, such as many machine learning algorithms (e.g., HaLoop, Spark). Lastly, there are projects such as Mesos, which aim to separate cluster management and scheduling concerns from the particulars of a given data-processing framework (e.g., MapReduce, general DAGs, iterative processing) and permit multiple such frameworks to coexist on the same cluster.

Cloud Systems

The computational demand of a particular user can be highly variable over time. Efforts to make more efficient use of resources include grid and cloud computing systems. These systems make a collection of resources available to a user and allow increases or decreases in resource allocation. Cloud computing can be attractive because the overhead of managing a large and complex system can be outsourced to specialists.

Amazon started offering cloud computing services in 2006. With its Elastic Compute Cloud, users can specify an operating system and application disk image to be loaded on virtual servers ranging from low-end to high-end. The Simple Storage Service (S3) provides rentable persistent storage. Large-scale distributed applications can run on the Amazon cloud. For example, Apache Hadoop is designed to use Elastic Compute Cloud servers accessing data stored in S3. The success of Amazon's cloud service has encouraged the development of other cloud computing offerings. For example, Microsoft offers the Azure cloud service, providing compute and persistent storage services similar to those provided by Amazon.

Parallel and Distributed Databases

A database is a system that provides facilities for reliably storing data and later retrieving it using a convenient program. For the purposes of this discussion, it is assumed that the database is relational—i.e., that its records consist of a particular set of fields each with a specific data type, and all records in a table have the same set of fields. It is further assumed that the database provides an SQL interface for accessing the data. Most commercial and open-source databases fit this description. These databases might also provide extensions for storing and querying semi-structured data (e.g., Extensible Markup Language, XML) and might support an extended query language; for example, one that supports recursive queries. However, these extensions are not necessary for this discussion.

Most commercial and open-source databases are parallelized in the sense that they can use multiple compute cores to evaluate a query when executed on a multicore server. One method of parallelization is to use multiple threads for performing expensive tasks, such as sorting or joining large data sets. Another method of parallelization takes advantage of the nature of the programs that a query in a language such as SQL will generate. An SQL query is converted into a collection of query operators connected into a rooted DAG (the query graph); the edges of the graph indicate data flow among the query operators. If the query operators operate in a pipelined fashion (continually accepting input and producing output), multiple query operators can execute in parallel, in a manner similar to the inter-operator parallelism exploited by data stream systems. Parallelizing database processing has been an active research topic for several decades.

Very-large-scale parallel database systems are generally spread over a collection of servers using a shared-nothing architecture—that is, there is no cluster file system to provide a shared state. The tables in a shared-nothing database are horizontally partitioned, and the partitions are distributed among the database servers. Each of the database servers can run a parallelized database in the sense of taking advantage of all available cores.

A table can be partitioned among the database servers in many ways. Two common choices are round-robin (new data are spread evenly among the servers) and hash (data are spread among the servers based on a hash of one or more fields of the table). Different tables can be partitioned using different techniques. Critical or frequently accessed tables can be stored two or more times using different partitioning for each copy. The servers in a shared-nothing database cooperate to evaluate a query. Recall that a query is transformed into a rooted DAG of query operators. The root of the DAG produces the query result, while the leaves access database tables. Subtrees of the query plan that operate on single tables are sent to the database servers, which compute the partial result represented by the subtree executing on the table partition local to the database server. The partial results generally need to be combined to form a result, whether for aggregation (which is similar to the reduce phase of a MapReduce program), or to join the result of the subtree with data from another subtree. This data transfer is represented by an operator commonly called shuffle. A complex query can involve a large query graph with many shuffle operators.

A shared-nothing parallel database can take many steps to optimize query evaluation. Some of the available techniques are as follows:

- Modify the way that one or more tables are partitioned among the database servers. For example, if tables R and S are frequently joined on key k, than one possible optimization is to partition both R and S on k, using the same hash function, to avoid a data shuffle when processing the join.
- If the result of two subtrees is to be joined, shuffle the results of the subtree that produces less data to the matching locations of the data from the other subtree result.
- Pipeline the operators to avoid the need to store very large partial results.

A database generally collects extensive statistics about the data that it manages and the queries that it processes to guide these and other optimizations. Shared-nothing databases were first developed in the 1980s (the Gamma and Grace research prototypes and the Teradata commercial DBMS), and extensive research has been performed on parallel database optimization. DeWitt and Stonebraker (2008) found that for data analysis tasks for which a relational database is well-suited, a shared-nothing relational database significantly outperforms a MapReduce program implemented using Hadoop. However, they also found that tuning parallel databases is often a difficult task requiring specialized expertise, whereas MapReduce systems are more readily configured to give good performance. Modern shared-nothing parallel databases include Teradata, Netezza, Greenplum, and extensions to Oracle and DB2.

However, a relational DBMS is not suitable for many analysis tasks. One well-known problem is that relational DBMSs are not well structured for managing array data—which are critical for many analyses. While the ability of modern databases to optimize storage layout and query evaluation plans makes array management with a database an attractive idea, the query optimization for array data is difficult, and the relational model is based on sets, not ordered data. Several efforts to incorporate array data into the relational model have appeared in the research literature, but without lasting effect. The open-source project SciDB is developing a parallel shared-nothing database system designed to support array data. This system supports parallelism by chunking large arrays and distributing them among the database servers.

A NoSQL database is loosely defined as being a data store that provides fewer consistency guarantees than a conventional database and/or a database that stores non-relational data, such as documents or graphs. NoSQL databases attempt to improve scaling by providing only weak or eventual consistency guarantees on the stored data, eliminating much of the complexity and overhead of the traditional strong consistency provided by conventional databases, which is especially marked in a distributed setting. Examples of NoSQL databases include MongoDB (document store), Neo4j (graphs), Bigtable, and Cassandra.

Parallel Programming Languages and Systems

Developing parallel and/or distributed programs is notoriously difficult, due to the problems in finding resources, distributing work, gathering results, recovering from failures, and understanding and avoiding rare conditions. A variety of tools have been developed to reduce the burden of developing parallel and distributed programs, for example, the Message Passing Interface and Remote Method Invocation in Java. However, parallel programming with these intermediate-level tools is still difficult because the programmer is forced to specify many details of how the parallelism is managed. Simple access to parallel programming seems to require languages that are at least partly functional (e.g., MapReduce) or declarative (e.g., SQL).

One method for achieving simple user parallelism is to create languages whose primitives perform expensive operations on very-large data structures. Large matrix operations, such as multiplication or inversion, are expensive but readily parallelizable. Languages such as Matlab, S, Splus, and R, for which the basic data structure is a matrix, may therefore be promising aids to parallelism. R is the open-source version of S-plus, and it has attracted the most development effort. Open-source efforts include Multicore-R and R/parallel, which add a parallelized Apply construct to the language. Revolution Analytics produces a commercial version of R with a variety of extensions to support large-scale data analysis, for example, external memory versions of commonly used statistical analyses, and parallelized versions of looping functions such as "foreach" and "apply," as well as multicore implementations of matrix operators. Ricardo interfaces R to Hadoop through Jaql and uses Jaql to run parallelized data analysis queries before loading the results into R for local analysis (Das et al., 2010).

Some new programming languages are designed to readily support parallel programming. For example, F# is a functional language derived from OCaml, but with simplified syntax. While F# supports an imperative programming style, its nature encourages a functional style.

TRENDS AND FUTURE RESEARCH

The clear trend for large-scale data analysis is to make increasing use of multicore parallel and distributed systems. The method for achieving enhanced performance through parallelism will depend on the nature of the data and the application. The largest analyses will be performed in large data centers running specialized software such as Hadoop over HDFS to harness thousands of cores to process data distributed throughout the cluster. Other large and centrally maintained facilities might run streaming analysis systems that reduce massive qualities of real-time data into a more manageable high-level data product.

However, actual analysts need to explore these data sets for them to be useful. One option is a grid-style environment in which users submit batch jobs to a large cluster and sometime later retrieve a result. While this result might be highly reduced, e.g., a plot in a graphical user interface, it might also be a processed data set delivered to the analysts' workstation (or cluster). Even inexpensive personal computers currently provide four high-performance cores and access to a powerful GPU. The local workstation will provide parallelized tools for exploring and analyzing the local data set.

While large server farms provide immense computing power, managing them is expensive and requires specialized technical expertise. Therefore the trend of outsourcing large computing tasks to cloud services such as Amazon's SC2 is likely to continue. One roadblock to using cloud services for massive data analysis is the problem of transferring the large data sets. Maintaining a high-capacity and wide-scale communications network is very expensive and only marginally profitable.

Software systems tend to develop greater power and performance until the complexity of the system exceeds human (or organizational) ability to manage it. Parallel databases allow naive users to compose and execute complex programs over petabytes of data. Similarly, MapReduce removes enough of the complexity of writing very-large-scale distributed programs that a large user group can access the power of a large cluster. Hadoop overlays further reduce the complexity of large-scale data analysis.

However, using and maintaining large parallel and distributed systems remains difficult for the following reasons:

- While parallel databases are readily queried by casual users, they are very hard to tune, and data loading remains a bottleneck.

- Although modern systems and languages have made parallel programming much easier than previously, they remain significantly more difficult than serial programs.
- While modern systems and languages abstract away many of the difficulties of parallel and distributed programming, debugging remains difficult.
- Architecting, building, and maintaining a large cluster requires specialized expertise.
- Understanding the performance of parallel and distributed programs and systems can be extremely difficult. Small changes to program phasing, data layout, and system configuration can have a very large effect on performance. Very large systems are typically accessed by a user community; the effect of the interaction of multiple parallel programs compounds the problem of understanding performance.

Achieving greater use of the power of parallel and distributed systems requires further innovations that simplify their use and maintenance. Many very-large data systems need to store very-large amounts of historical data, but also provide real-time or near-real-time alerting and analytics. However, systems designed for real-time response tend to have a very different architecture than historical, batch-oriented large-data systems. A typical response to these needs is to build two separate and loosely coupled systems. For example, a streaming system might provide real-time alerting, while historical analyses are made on a batch-oriented system. Transparently bridging real-time systems with large-data systems remains a research issue.

Similarly, data integration and data-quality assurance are difficult problems, which, in spite of tools such as Clio (Haas et al., 2005) or IBM's InfoSphere Information Analyzer, are generally bespoke, labor-intensive tasks. A significant direction of future research is the development of simple but powerful data-integration and data-quality tools that use machine learning techniques to automate these tasks.

REFERENCES

Asanovic, K., R. Bodik, B.C. Catanzaro, J.J. Gebis, P. Husbands, K. Keutzer, D.A. Patterson, W.L. Plishker, J. Shalf, S.W. Williams, and K.A. Yelick. 2006. *The Landscape of Parallel Computing Research: A View from Berkeley.* University of California, Berkeley, Technical Report No. UCB/EECS-2006-183. December 18. Available at http://www.eecs.berkeley.edu/Pubs/TechRpts/2006/EECS-2006-183.html.

Balakrishnan, M., D. Malkhi, V. Prabhakaran, T. Wobber, M. Wei, and J.D. Davis. 2012. CORFU: A shared log design for flash clusters. Pp. 1-14 in *9th USENIX Symposium on Networked Systems Design and Implementation.* USENIX Association, Berkeley, Calif., April.

Das, S., Y. Sismanis, S. Beyer, R. Gemulla, P.J. Haas, and J. McPherson. 2010. Ricardo: Integrating R and Hadoop. Pp. 987-998 *Proceedings of the 2010 ACM SIGMOD International Conference on Management of Data*. Association for Computing Machinery, New York, N.Y.

Dean, J., and S. Ghemawat. 2004. MapReduce: Simplified data processing on large clusters. P. 10 in *Proceedings of the 6th Symposium on Operating Systems Design and Implementation*. USENIX Association, Berkeley, Calif.

DeWitt, D.J., and M. Stonebraker. 2008. "MapReduce: A Major Step Backwards." Online discussion. Available at http://homes.cs.washington.edu/~billhowe/mapreduce_a_major_step_backwards.html.

Ghemawat, S., H. Gobioff, and S.-T. Leung. 2003. The Google file system. Pp. 29-43 in *Proceedings of the Nineteenth ACM Symposium on Operating System Principles*. Association for Computing Machinery, New York, N.Y.

Haas, L.M., M.A. Hernández, L. Popa, M. Roth, and H. Ho. 2005. Clio grows up: From research prototype to industrial tool. Pp. 805-810 in *Proceedings of the 2005 ACM International Conference on Management of Data (SIGMOD)*. Association for Computing Machinery, New York, N.Y.

4

Temporal Data and
Real-Time Algorithms

INTRODUCTION

Temporal data are associated with real-time acquisition and prediction of either human-generated data (e.g., Web traffic) or physical measurements (e.g., speech and video data). Temporal information sources are very relevant to the challenges of analyzing massive data because many massive streams of data exhibit real-time properties. Thus, real-time algorithms for managing temporal streams comprise a broadly useful foundation on which to create new analysis capabilities. This chapter focuses on the current solutions for and the specific challenges that time imposes on tasks such as data acquisition, processing, representation, and inference. It illuminates the challenges of dynamic data, and it will also touch on the hardware infrastructure required for storing and processing temporal data.

An example of the changes wrought by time upon massive data sets for human-generated data is the "click-through rate" estimation problem in online advertising systems. Millions of new data elements are accumulated every day, and the number of dimensions (number of discrete click-through paths) may grow by a few thousand per day. However, old dimensions, also referred to as coordinates, also disappear, such as when a proper noun that was frequent in the past is no longer used. For example, a new word like "iPad" adds a dimension, while a specific typewriter that is no longer manufactured may disappear from the relevant data, eliminating a dimension. Natural sequences such as speech and audio signals exhibit similar characteristics, although the dimension does not grow as rapidly. A notable example here is speech excerpts collected from mobile devices. Here the sheer number of utterances, their variability (e.g., accents and dialects), and the vocabulary size pose serious challenges in terms of storage, representation, and modeling. Last, but not least, is the domain of real-time imaging streams from satellites, surveillance cameras, street-view cameras, and automated navigation machines (such as unmanned cars and small aerial surveillance vehicles), whose collective data is growing exponentially.

DATA ACQUISITION

The initial phase of a temporal data analysis system is the acquisition stage. While in some cases the data are collected and analyzed in one location, many systems consist of a low-level distributed acquisition mechanism. The data from the distributed sources must generally be collected into one or more data analysis centers using a real-time, reliable data feeds management system. Such systems use logging to ensure that all data get delivered, triggers to ensure timely data delivery and ingestion, and intelligent scheduling for efficient processing. For social media, data are often analyzed as they are collected, and the raw data are often not archived due to lack of storage space and usage policies.

Real-time massive data analysis systems generally use some type of *eventual consistency,* which, as the term implies, means that eventually the data arrive to all servers. Eventual consistency is often used in large-scale distributed systems to minimize the cost of distributed synchronization. Eventual consistency is also appropriate for real-time data analysis, because generally one does not know when all relevant data have arrived. Failures and reconfigurations are common in very-large-scale monitoring systems, so, in general, one cannot determine whether a data item is missing or merely late. Instead, the best strategy is generally to do as much processing as possible with the data that are available, and perhaps recompute answers as additional data come in.

Large-scale real-time analysis systems not only collect a data stream from many sources, they also typically collect many data streams and correlate their results to compute answers. Different data streams typically are collected from different sources, and they often use different data-feed delivery mechanisms. As a result, different data streams typically exhibit different temporal latencies—one might reflect data within 1 minute of the current time, another within 10 minutes of the current time. Differing latencies in data streams, combined with the uncertainty associated with determining when all data up to time t for a stream have been collected, make it difficult to produce definitive results for a query without a significant delay. The problem of determining when a collection of data streams can produce a sufficiently trustworthy answer up to time t is called *temporal consistency*. The theory and practice of temporal consistency of streams is at its infancy.[1]

Large real-time data analysis systems will often collect many real-time data streams and compute many higher-level data products (*materialized views*) from them. Many data-ingest and view-update tasks must compete for limited system resources. Conventional real-time scheduling theories (such as hard, firm, or soft real-time scheduling) are not appropriate, because tasks that miss deadlines either break the system (hard real-time), are discarded (firm real-time), or are ignored (soft real-time). The recent theory of *bounded-tardiness* scheduling (Leontyev and Anderson, 2010) provides the most appropriate way to model a real-time data analysis system. Tasks can miss their deadline without breaking the system or being discarded, but their *tardiness* in completion after their deadline is bounded. Most conventional real-time scheduling algorithms, such as earliest-deadline first, are bounded-tardiness algorithms.

Massive real-time data warehouses also need to cope with the breakage of one or more temporal feeds. Such a breakage might be the failure of a server at the feed side, an unannounced change in schema, and so on. When the feed source recovers, its past stream needs to be ingested and all transitively dependent data products updated. This task places a huge load on a temporal massive data warehouse, throwing it into temporary overload and creating the need for a graceful recovery of the affected tables without degrading the timeliness of updates to the other tables in the warehouse. The problem becomes more pronounced when a stream warehouse system needs to store a long history of events, e.g., years or decades, and is continuously loaded. Moreover, such stream warehouses also need to cope with the need of providing both immediate time alerts and long-range aggregated statistics. There is thus a tension in such systems between timely serving needs and the synchronization latency, which is necessary for maintaining consistency.

In some online transaction processing systems, fast real-time synchronization can become an issue (Kopetz, 1997). When data integrity is a mandatory requirement, the state-of-the art systems use some variation of the Paxos algorithm. Paxos is actually a family of protocols for determining consensus in a network of unreliable processors; consensus is the process of agreeing on the result among a group of computing units, which is difficult when the units or their communication medium experience temporal failures. However, the Paxos family of algorithms was designed for maintaining consistency in small- to medium-scale distributed data warehousing systems, and scaling Paxos-based and other consistency preserving storage

---

[1] For an initial keynote paper that suggests a formal treatment of stream consistency, see Golab and Johnson (2011).

mechanisms is currently a critical open issue. In practice, implementing a consistent distributed real-time system in a massive computing environment with frequent transactions requires special technical expertise. As described in the President's Council of Advisors on Science and Technology report *Designing a Digital Future* (PCAST, 2010), the challenge in building large-scale temporal systems is that they must be robust to hardware failures as well as software bugs. For example, because a modern central processing unit (CPU) has a failure rate of about one fatal failure in 3 years, a cluster of 10,000 CPUs would be expected to experience a failure every 15 minutes on average. A temporal system for massive data must maintain a consistent, temporally coherent view in spite of this. As a real-world example, the Paxos algorithm lies at the heart of Google's cluster servers for real-time transactions and services. Despite the fact that the Paxos algorithm, which was invented more than 20 years ago, is well understood and analyzed, a 1-year effort by one of the world's experts in the field of distributed processing was still necessary to implement the algorithm on Google's cluster system at a speed that will sustain the required transaction rate as well as survive a burst of failures.[2]

DATA PROCESSING, REPRESENTATION, AND INFERENCE

The next stage in time-aware data analysis includes building an abstract representation of the data and then using it for inference. Methods for abstract data representation include coding and sketching. The coding sub-phase is based on either perceptual codes, which are often lossy, or lossless source coding techniques. Lossless source codings are naturally suitable for encoding temporal streams because they are often based on Markov models, efficiently represented as a context tree.[3] The context modeling is combined with a backend stage, which is based on arithmetic coding. Coding systems are typically designed under the assumption that only a constant space is available. While this assumption is asymptotically valid, recent advances in flash-based memory architectures may greatly enhance the current state-of-the-art algorithms.

To cope with the computational needs that real-time and long-range temporal queries impose, analytic tools for summarizing temporal data streams are a must. A common and very effective summarization tool is called sketching. There are several types of sketching representations, broken down into two broad categories: those that retain data in its native format (e.g., sliding windows, a technique for randomly sub-sampling time

---

[2] For further details see Chandra et al. (2007).

[3] See Eindhoven University of Technology, The Context-Tree Weighting Project, available at http://www.sps.ele.tue.nl/members/F.M.J.Willems/RESEARCH_files/CTW/ResearchCTW.htm.

series) and those that use some derived format (e.g., random projections of one or more data coordinates, histograms of underlying distribution). Combinations of these two types of representation are also used. Chapter 6 contains more discussion of sketching.

Many data sources have an inherent periodic component or a natural time scale, and natural time-aware representations include averaged snap-shots or windows of data over time, e.g., averaged over every basic time scale (such as a day) or repeated for many periods of time. See the tutorial by Garofalakis et al.[4] for a number of examples of representation types and techniques. One key mathematical feature (albeit not a necessary feature) of any time-aware representation method is that it be linear; the representation of changes over time in the data are easily reflected and easily computed in changes to the original representation.

Going past the representation phase, which can be the sole stage of a real-time system, the core of many temporal data streams is a learning and inference engine. There has been an immense amount of work on online algorithms that are naturally suitable for time-aware systems.[5] Most on-line algorithms impose constant or at least sublinear memory assumptions, similar to data-streams algorithms. However, to cope with non-stationarity effects (changes in the distribution of the input stream) and to achieve high accuracy, more computationally demanding and space-consuming ap-proaches are needed. One notable and promising approach is mixed online and batch learning by follow-the-regularized-leader (FTRL) algorithms, an overview of which is given in the book by Cesa-Bianchi and Lugosi.[6] To date, however, there have been few implementations of large-scale massive data analysis systems based on FTRL.

In addition to the use of temporal data to form accurate predictions, the processing of temporal data often gives rise to specialized inference problems such as change-point detection. When the input data rate exceeds the computing capabilities of online learning and prediction algorithms, one needs to resort to methods that provide approximate representations. This paradigm is often referred to as the data-stream approach. Data-stream algorithms provide temporal tools for representing and processing input data that come at a very high rate. The high-rate input stresses the com-munication, storage, and computing infrastructure to the point that it is difficult, if not impossible, to transmit the entire input, compute complex functions over large portions of the input stream, and store and capture

---

[4] M. Garofalakis, J. Gehrke, and R. Rastogi, "Querying and Mining Data Streams: You Only Get One Look. A Tutorial," presented at the 28th International Conference on Very Large Data Bases (VLDB 2002), August 20-23, 2002, available at http://www.cse.ust.hk/vldb2002/program-info/tutorial-slides/T5garofalalis.pdf, accessed June 16, 2012.

[5] Cesa-Bianchi and Lugosi (2006) provides a broad in-depth description of online algorithms.

[6] Ibid.

temporally the entire input stream. Numerous effective fast algorithms exist for extracting statistical quantities such as median, mean, quantiles, and histograms and, more generally, for answering queries of the data set or multiple data sets. In the past 10 years, there have been a number of stream-based data management systems developed to address these questions. One such example is the Stanford Data Stream Management System. Theory and applications of streaming data have developed to the point where a whole book has been dedicated to the subject (Muthukrishnan, 2005). However, the fusion of stream approaches with efficient statistical inference for general models remains a major research challenge. This fusion poses significant challenges because state-of-the-art learning algorithms are not designed to cope with partial summaries and snapshots of temporal data.

SYSTEM AND HARDWARE FOR TEMPORAL DATA SETS

The discussion thus far has focused on software, analysis, and algorithmic issues and challenges that are common to massive temporal data. Massive temporal data also pose high demands on the hardware and systems infrastructure. Such systems need to employ a very large distributed file system such as Google's file system (GFS) and tens of data-acquisition machines to funnel the data to thousands of processors using very fast interconnects. This type of architecture has a very high throughput but is very difficult to replicate and expensive to maintain, requiring a good complement of reliability engineers. Massive temporal systems cannot be deployed by boutique-size data warehouses because there are only a handful of tools that can help in large-scale processing. Noise-tolerant storage of temporal data also places a high bar on maintaining data integrity because storage error patterns tend to be local and bursty in nature. Although the theory of error correction for communication over channels prone to burst errors is well established (e.g., McAuley, 1990), applications of the theory to massive storage of temporal data are mostly confined to proprietary systems such as the aforementioned GFS. In addition to the storage requirements, substantial computing infrastructure is required even for simple tasks. Here again there is a lack of publicly available source code for near-real-time processing of temporally stored data.

CHALLENGES

Major current and future challenges that arise in time-aware systems for massive data include the following:

- *Design and implementation of new representation algorithms and methods for perpetually growing, non-stationary massive data, especially in conjunction with learning and modeling.* Although sketching algorithms for streaming data naturally incorporate changes in the data streams, they do not necessarily give an easy and straightforward method for adjusting and updating models and inferences derived from these sketches over time. Current algorithms permit efficient model-building but do not efficiently change the models over time. Furthermore, there is not a natural way to identify or to detect model changes in a streaming setting, perhaps with limited data. The current algorithms for updating network metrics permit efficient calculation only for certain network structures.

- *Streaming and sketching algorithms that leverage new architectures, such as flash memory and terascale storage devices.* As discussed in the chapter on sampling, software and hardware models for acquiring data quickly over time is an area of active current research. Many streaming and sketching algorithms are designed in the absence of a specific hardware or software system; yet it is only when practical systems are built that both the limitations of the theoretical algorithms as well as potential new algorithms are seen.
- *Distributed real-time acquisition, storage, and transmission of temporal data.*
- *Consistency of data.* Most systems perform acquisition in an asynchronous manner. When consistency is important, Paxos-based algorithms are employed. Can these solutions scale when the input stream is one or two orders of magnitude more massive, as in the case of audio and video data?
- *Lack of effective tools for the design, analysis, implementation, and maintenance of real-time, temporal, time-aware systems for nonprofit, educational, and research institutions, including lack of realistic data sources for benchmarking algorithms and hardware performance.*

REFERENCES

Cesa-Bianchi, N., and G. Lugosi. 2006. *Prediction, Learning and Games.* Cambridge University Press, New York, N.Y.

Chandra, T., R. Griesemer, and J. Redstone. 2007. Paxos made live—An engineering perspective. *PODC '07: 26th ACM Symposium on Principles of Distributed Computing.* Available at http://labs.google.com/papers/paxos_made_live.html.

Golab, L., and T. Johnson. 2011. Consistency in a stream warehouse. Pp. 114-122 in *Proceedings of the 2011 Conference on Innovative Data Systems Research (CIDR).* Available at http://www.cidrdb.org/cidr2011/program.html.

Kopetz, H. 1997. *Real-Time Systems: Design Principles for Distributed Embedded Applications*. Kluwer Academic Publishers, Norwell, Mass.

Leontyev, H., and J.H. Anderson. 2010. Generalized tardiness bounds for global multiprocessor scheduling. *Real-Time Systems* 44(1):26-71.

McAuley, A.J. 1990. Reliable broadband communication using burst erasure error correcting code. *ACM SIGCOMM Computer Communication Review* 20(4):297-306.

Muthukrishnan, S. 2005. *Data Streams: Algorithms and Applications*. Now Publishers, Hanover, Mass.

PCAST (President's Council of Advisors on Science and Technology). 2010. *Designing a Digital Future: Federally Funded Research and Development in Networking and Information Technology*. Office of Science and Technology Policy, Washington, D.C.

5

Large-Scale Data Representations

OVERVIEW

Data representation refers to the choice of a mathematical structure with which to model the data or, relatedly, to the implementation of that structure. The choice of a particular data representation is dictated by various considerations, such as hardware, communication, data-generation process, input-output, data sparsity, and noise. As such, questions of large-scale data representation typically have algorithmic, statistical, and implementation or systems aspects that are intertwined and that need to be considered jointly. In addition, the nature of data representation changes depending on the task at hand, the context in which the data are acquired, the aspect of data analysis being addressed, and what features of the data are being captured.

A closely related but somewhat more vague notion is that of a data feature. In many cases, a data feature is an externally defined property of the data that can be easily computed from the data or measured directly and then plugged into a data-processing algorithm. Designing such features is often difficult, because it requires substantial domain-specific insight, but it is often the most important step in the data-analysis pipeline. In other cases, however, standard mathematical operations are used to transform the data and, thereby, define features. For example, properties of the eigenvectors or eigenvalues of matrices associated with the data can be used as features.

For consistency, the following taxonomy is used to distinguish between several basic uses of the term "data representation":

- *Basic data structures.* This category includes structures such as hash tables, inverted indices, tables/relations, etc. These are data structures that one finds in standard textbooks on algorithms and databases (Cormen et al., 2009; Garcia-Molina et al., 2008).
- *More abstract, but basic, mathematical structures.* This category includes structures such as sets, vectors, matrices, graphs, and metric spaces. Many of these mathematical structures have sparse variants, e.g., vectors and graphs with few non-trivial components, matrices of low rank, and so on. Each mathematical structure can be represented using many data structures, with different implementations supporting different operations and optimizing different metrics.
- *Derived mathematical structures.* This category includes more sophisticated structures such as clusters, linear projections, data samples, and such. Although these representations are often basic mathematical structures, they do not directly represent the original data, but are instead derived from the data and often can be viewed as representations of components in a model of the data.

The committee also differentiates between two different "design goals" that one has to keep in mind in choosing appropriate data representations:

- First, on the upstream or data-acquisition or data-generation side, one would like a structure that is "sufficiently close" to the data. Such a representation supports the process of generating, storing, or accessing the data while providing a model for the noise or uncertainty properties of the data. The goal of such representations is to support all of these operations without significantly altering the data.
- Second, on the downstream or data analysis side, one would like a structure that has both a flexible description and is tractable algorithmically. That is, it should be expressive enough so that it can describe a range of types of data, but it should not be so expressive that it can describe anything (in which case computations and inferences of interest would likely be intractable).

GOALS OF DATA REPRESENTATION

Although a picture may be worth a thousand words, a good representation of data is priceless: a single data representation, or sometimes multiple ones, allows one to carry out a large number of data processing and analysis tasks in a manner that is both algorithmically efficient and statistically meaningful. This section presents an overview of the various goals of data representations, together with illustrative examples of representations that accomplish those goals.

Reducing Computation

In the massive data context, a key goal in data representation is that of reducing computation. In order to understand the relationship between data representation and computational efficiency, one has to examine specific data structures that are most suitable for different computational primitives.

One basic operation in data processing is to query parts of the data. The main technique to facilitate efficient querying is indexing, and thus one needs advanced indexing algorithms that can retrieve desired data and attributes efficiently. The indices can be regarded as additional representations of the data that are derived from the raw data and added to the data set as auxiliary information. Moreover, different classes of queries and different data require different indexing methods in order to be performed efficiently.

For example, for queries on text documents, a standard indexing technique is an inverted index table (Baeza-Yates and Ribeiro-Neto, 1999). That is, one stores a mapping from content (such as words or numbers) to its physical locations in the data set. One may map different contents into one index; for example, an indexer for text search usually records words with the same underlying meaning (e.g., "drives," "drove," and "driven" would be related to the single concept word "drive"). In some applications (e.g., biology) it is important to allow more general pattern matching queries, where the goal is to find arbitrary substrings (not only words) in the text. Suffix trees, suffix arrays, and variations of these are some of the popular structures solving the pattern matching problem (Jones and Pevzner, 2004). In general, approximate matching may require more complex indexing schemes. Example are locality-sensitive hashing (Andoni and Indyk, 2008) and min-hashing (Broder, 1997), which can efficiently find approximate nearest neighbors of a query from a database.

For many other aspects of data analysis, the more traditional data representations have been geared toward algorithms that process data sequentially on a single machine. Additional considerations are needed for large-scale distributed computing environments that use computational resources from multiple computers.

Reducing Storage and/or Communication

In addition to reducing computation time, proper data representations can also reduce the amount of required storage (which translates into reduced communication if the data are transmitted over a network). For example, text documents can be efficiently compressed using Lempel-Ziv compression, the Burrows-Wheeler transform, or similar representations (Salomon, 1998). Compression can be also applied to data structures, not just data. For the aforementioned pattern-matching search problem, there exist data structures (based on the Burrows-Wheeler transform) that store only a compressed representation of the text (Ferragina and Manzini, 2000; Langmead et al., 2009).

One can reduce the storage size even further by resorting to lossy compression. There are numerous ways this can be performed. This is often achieved by transform coding: one transforms the data (e.g., images or video) into an appropriate representation (e.g., using Fourier or wavelet transforms) and then drops "small" coefficients. For matrices, lossy compression can be achieved by truncating the Singular Value Decomposition or the eigenvalue decomposition (Hansen, 1987). This involves retaining just the dominant part of the spectrum of the data matrix and removing the "noise." Different forms of lossy compression can be used to improve space efficiency of data structures as well. Examples include Bloom filters, for approximate set-membership queries (Broder and Mitzenmacher, 2002), and Count-Sketch (Charikar et al., 2002) and Count-Min (Cormode and Muthukrishnan, 2005) representations for estimating item counts. These methods randomly map (or hash) each data item into a lower-dimensional representation, and they can be used for other tasks, such as estimating similarity between data sets. Other related techniques for obtaining low-dimensional representations include random projections, based either on the Johnson-Lindenstrauss lemma (Johnson and Lindenstrauss, 1984) or its more time-efficient versions (Ailon and Chazelle, 2010; Ailon and Liberty, 2011).

Reducing Statistical Complexity and Discovering the Structure in the Data

One can also use data representations to reduce the statistical complexity of the problem—the amount of data needed to solve a given statistical task with a given level of confidence—by approximating the data set by simpler structures. Specifically, by carefully selecting a small number of data attributes, data points, or other parameters, one can apply more constrained statistical models with fewer parameters and improve the quality of statistical inference. The reduced representation should represent the sufficient statistics for inference, while the "dropped" data should either be irrelevant or simply be noise. Alternatively, one can view this process from a model-based perspective as discovering the latent structure of the data.

Typically, this approach also leads to reduced running time and storage. This can be exploited as a filtering approach to data analysis: if we have a data set with many features that cannot be processed by complex algorithms, we may use a simple feature-selection algorithm to filter the features and obtain a smaller (but still reasonably large) set of features that can be handled by the complex algorithm. This leads to a multi-stage approach that goes back and forth between representations with different accuracies and algorithmic complexity.

Two basic approaches to structure discovery are dimensionality reduction and clustering.

Dimensionality reduction refers to a broad class of methods that re-express the data, typically in terms of vectors that are formally of very high dimension, in terms of a small number of actual data points or attributes, linear or nonlinear combinations of actual data points/attributes, or linear combinations of nonlinearly transformed actual data points/attributes. Such methods are most useful when one can view the data as a perturbed approximation of some low-dimensional scaffolding. There are numerous algorithms for discovering such structures in data, either linear (e.g., Principal Component Analysis; see van der Maaten et al., 2009) or nonlinear ones (locally linear embedding, isomap, Laplacian eigenmaps, diffusion maps; see Roweis and Saul, 2000; Tenenbaum et al., 2000; Belkin and Niyogi, 2003; and Coifman and Lafon, 2006).

It is critically important to understand what kind of properties must be preserved via dimensionality reduction. In the simplest unsupervised settings, it is often the variance or some other coarse measure of the information in the data that must be preserved. In supervised settings, the goal is to reduce the dimension while preserving information about the relevant classification directions. A popular method for achieving this goal when it is thought that some sort of sparsity is present is to augment a regression or classification objective with an L_1 constraint, which reduces the number of attributes used by the classifier (Tibshirani, 1996). A different class of methods, known as sufficient dimension reduction methods, search over projections of the data to find a projection that retains as much information about the response variable as possible.

Although algorithms that implicitly incorporate sparsity (e.g., local spectral methods or local diffusion-based or sampling-based methods) have been less well-studied than algorithms that explicitly add a regularization term to the original objective, the former have clear advantages for very-large-scale applications. Relatedly, randomization as a computational or algorithmic resource, and its implicit regularization properties, is also a subject of interest, in particular for large-scale data analysis. By implicitly incorporating regularization into the steps of an approximation algorithm in a principled manner, one can obtain algorithmic and statistical benefits simultaneously. More generally, although the objects in the reduced dimensional space are often represented as vectors, other useful data representations such as relational or hierarchical representations can be useful for large-scale data that are not naturally modeled as matrices.

Dimensionality reduction is often used after increasing the dimensionality of the data. Although at first this may appear counter-intuitive, the additional dimensions may represent important nonlinear feature interactions in the data. Including these additional dimensions can simplify statistical inference because the interaction information they capture can be incorporated into simpler linear statistical models, which are easier to deal with. This forms the basis of kernel-based methods that have been popular recently in machine learning (Hofmann et al., 2008), and, when coupled with matrix sampling methods, forms the basis of the Nystrom method (Kumar et al., 2009).

Clustering (also known as *partitioning* or *segmentation*) is a widely performed procedure that tries to partition the data into "natural groups." Representing the data by a small number of clusters necessarily loses certain fine details, but it achieves simplification, which can have algorithmic and statistical benefits. From a machine learning perspective, this task often falls under the domain of unsupervised learning, and in applications often the first question that is asked is, How do the data cluster? Many clustering algorithms are known in the literature. Some of the most widely used are hierarchical clustering, k-means, and expectation maximization (Jain et al., 1999).

There is a natural relationship between clustering and dimensionality reduction methods. Both classes of methods try to make data more compact and reveal underlying structure. Moreover, one may view clustering as a form of dimensionality reduction. Some methods, e.g., latent Dirichlet allocation (Blei et al., 2003), explicitly exploit this connection.

Exploratory Data Analysis and Data Interpretation

Both dimensionality reduction and clustering are highly useful tools for visualization or other tasks that aid in initial understanding and interpretation of the data. The resulting insights can be incorporated or refined in more sophisticated inference procedures. More generally, exploratory data analysis refers to the process of simple preliminary examinations of the data in order to gain insight about its properties in order to help formulate hypotheses about the data. One might compute very simple statistics, use some of the more sophisticated algorithms discussed earlier, or use visualization tools. Sampling the data, either randomly or nonrandomly, is often important here, because working interactively with manageable sizes of data can often be a first step at determining what is appropriate for a more thorough analysis, or whether a more sophisticated analysis is even needed. It should be noted, however, that in addition to reducing storage, sampling methods such as CUR decompositions[1] can also be also useful for exploratory data analysis tasks, because they provide actual data elements that are representative in some sense. For example, when applied to genetics data, CUR decompositions can provide actual information about actual patients or actual DNA single-nucleotide polymorphisms that can then be typed in a laboratory by a geneticist for further downstream analysis (Paschou et al., 2007).

Sampling and Large-Scale Data Representation

Many questions of large-scale data representation have to do with questions of sampling the data, and there are several complementary perspectives one can adopt here.

On the one hand, one can imagine that the data have been collected (either actually or conceptually), and one is interested in performing relatively expensive computations on the data. One possible solution, mentioned above, is to perform the expensive computation on a small sample of the data and use that as an approximation for the exact solution on the full data set. In this case, sampling is effectively a dimensionality-reduction method and/or a technique for reducing storage that makes it possible to obtain approximate solutions to expensive problems. Many randomized, as well as nonrandom, sampling techniques are known as well, including core-sets, data squashing, and CUR decompositions (Agarwal et al., 2005; DuMouchel, 2002; Mahoney and Drineas, 2009). More generally, data clustering (also mentioned above) can be used to perform data compression by replacing each cluster element with a cluster "representative," a process sometimes known as vector quantization (Gersho and Gray, 1992).

On the other hand, sampling can refer to the process of collecting data in the first place, and typically it is of interest when collecting or analyzing all of the data is unreasonable. A more detailed discussion of this is provided in Chapter 8, and here the committee simply notes some similarities and differences. When sampling to collect data, one is interested in sampling enough to reduce the uncertainty or variance of, for example, estimates, but a major concern is also on controlling the bias. That is, one wants to collect at least some data from all groups or partitions that are thought to be relevant to the outcomes of the data analysis. When sampling is used to speed up computation, one is also interested in minimizing variance, but controlling bias is often less of an issue. The reason is that one can typically construct estimators that are unbiased or that have a small bias and then focus on collecting enough samples in such a way that the

---

[1] The decomposition of a data matrix into the factors "C," "U," and "R" is explained in Mahoney and Drineas (2009).

variance in the computed quantities is made sufficiently small. Because the goal is simply to speed up computations, one is interested in not introducing too much error relative to the exact answer that is provided by the larger data set. Thus, many of the underlying methodological issues are similar, but in applications these two different uses of the term "sampling" can be quite different.

CHALLENGES AND FUTURE DIRECTIONS

The discussion to this point has addressed some of the challenges in data representation in general, but it has not highlighted the additional burdens that massive data impose on data representations. The following subsections address key issues in representing data at massive scales. The challenges of data management per se are important here, but they have already been discussed in Chapter 3.

The Challenge of Architecture and Algorithms: How to Extend Existing Methods to Massive Data Systems

A large body of work currently exists for small-scale to medium-scale data analysis and machine learning, but much of this work is currently difficult or impossible to use for very-large-scale data because it does not interface well with existing large-scale systems and architectures, such as multicore processors or distributed clusters of commodity machines. Thus, a major challenge in large-scale data representation is to extend work that has been developed in the context of single machines and medium-scale data to be applicable to parallel, distributed processing and much larger-scale situations. A critical concern in this regard is taking into account real-world systems and architectural constraints. Many of these issues are very different than in the past because of the newer architectures that are emerging (Asanovic et al., 2006).

In order to modify and extend existing data analysis methods (that have implicitly been developed for serial machines with all the data stored in random access memory), a critical step will be choosing the appropriate data representation. One reason is that the choice of representation affects how well one can decompose the data set into smaller components so that analysis can be performed independently on each component. For example, intermediate computational results have to be communicated across different computational nodes, and thus communication cost should be minimized, which also affects and is affected by the data representation. This issue can be rather complicated, even for relatively simple data structures such as matrix representations. For example, to perform matrix-vector multiplication (or one of many linear algebra operations that use that step as a primitive), should the matrix be distributed by columns (in which case each node keeps a small set of columns) or should it be distributed by rows (each node keeps a small set of rows)? The choice of representation in a parallel or distributed computing environment can heavily impact the communication cost, and thus it has far-reaching consequences on the underlying algorithms.

The Challenge of Heavy-Tailed and High-Variance Data

Although dimensionality reduction methods and related clustering and compact representations represent a large and active area of research with algorithmic and statistical implications, it is worth understanding their limitations. At a high level, these methods take advantage of the idea that if the data are formally high-dimensional but are very well-modeled by a low-dimensional structure—that is, the data are approximately sparse in some sense—then a small number of coordinates should suffice to describe the data. On the other hand, if the data are truly high-dimensional, in the sense that a high-dimensional structure provides the best null model with which to explain the data, then measure concentration and other intrinsically high-dimensional phenomena can occur. In this case, surprisingly, dimensionality reduction methods can often be appropriate again. Both very-high-dimensional and very-low-dimensional extremes can have beneficial consequences for data analysis algorithms.

Unfortunately, not all data can fruitfully be approximated by one of those two limiting states. This is often the case when the processes that underlie the generation of the data are highly variable. Empirically, this situation often manifests itself by so-called heavy-tailed distributions, for example, degree distribution or other statistics of informatics graphs that decay much more slowly than do the tails of Gaussian distributions (Clauset et al., 2009).

As an example of this, social and information networks often have very substantial long-tailed or heavy-tailed behavior. In this case, there are a small number of components that are "most important," but those components do not capture most of the information in the data. In such cases, data often cluster poorly—there may be local pockets of structure, but the data might be relatively unorganized when viewed at larger size scales—and thus the vast majority of traditional algorithmic and statistical methods have substantial limitations in extracting insight (Leskovec et al., 2009).

Far from being a pathological situation, the phenomenon of data having good small-scale structure but lacking good large-scale structure arises in many applications. In Internet applications, a major challenge is to "push mass out to the heavy tail." For example, in a recommender system, the "heavy tail" may refer to the large number of users who rate very few movies, in which case a challenge is to make useful personalized recommendations to those users; or it may refer to the large number of movies that are not blockbuster hits and that have received very few ratings, in which case a challenge is to create new markets by suggesting these movies to potentially interested users. In both cases, there is substantial competitive advantage to be exploited by making even slightly more reliable predictions in this noisy regime. Similarly, in many scientific applications, the "heavy tail" is often where new scientific phenomena manifest themselves, and thus methods for extracting small signal from a background of noise is of interest.

Thus, a major challenge in large-scale data representation is to develop methods to deal with very-high-variance situations and heavy-tailed data.

The Challenge of Primitives: Develop a Middleware for Large-Scale Graph Analytics

From the computer systems perspective, it would be very helpful to identify a set of primitive algorithmic tools that (1) provide a framework to express concisely a broad scope of computations; (2) allow programming at the appropriate level of abstraction; and (3) are applicable over a wide range of platforms, hiding architecture-specific details from the users. To give an example of such a framework, linear algebra has historically served as such middleware in scientific computing. The mathematical tools, interactive environments, and high-quality software libraries for a set of common linear algebra operations served as a bridge between the theory of continuous physical modeling and the practice of high-performance hardware implementations. A major challenge in large-scale data representation is to develop an analogous middleware for large-scale computations in general and for large-scale graph analytics in particular. The Graph 500 effort[2] may be helpful in this regard, and Chapter 10 of this report discusses a possible classification of analysis tasks that might underpin a library of such middleware.

A good example of an archetypal problem for this challenge is the need for methods appropriate for the analysis of large but relatively unstructured graphs, e.g., on social networks, biological networks, certain types of noisy graphical models, etc. These problems represent a large and growing domain of applications, but fundamental difficulties limit the use of traditional data representations in large-scale applications. For example, these problems are characterized by poor locality of reference and data access, nontraditional local-global coupling of information, extreme sparsity, noise difficulties, and so on.

---

[2] See The Graph 500 List, available at http://www.graph500.org/.

Of particular importance is the need for methods to be appropriate for both next-user-interaction applications and interpretable analytics applications. In many applications, there is a trade-off between algorithms that perform better at one particular prediction task and algorithms that are more understandable or interpretable. An example of the former might be in Internet advertising, where one may wish to predict a user's next behavior. The most efficient data-analysis algorithm might not be one that is interpretable by a human analyst. Examples of interpretable analytics applications arise in fields such as biology and national security, where the output of the data-analysis algorithm is geared toward helping an analyst interpret it in light of domain-specific knowledge. Ideally, middleware would be designed to serve both of these purposes.

One of the potential approaches to this challenge is to use the existing middleware for linear-algebraic computation, using deep theoretical connections between graph theory and linear algebra (both classic (e.g., Chung, 1992) and more recent ones (e.g., Spielman, 2010). However, it is not clear yet to what extent these connections can be exploited practically to create an analogous middleware for very-large-scale analytics on graphs and other discrete data. Because the issues that arise in modern massive data applications of matrix and graph algorithms are very different than those in traditional numerical linear algebra and graph theory—e.g., the sparsity and noise properties are very different, as are considerations with respect to communication and input/output cost models—a central challenge will be to deal with those issues.

The Challenge of Manipulation and Integration of Heterogeneous Data

The manipulation and integration of heterogeneous data from different sources into a meaningful common representation is a major challenge. For example, in biology, one might have mRNA expression data, protein-protein interaction data, gene/protein sequence data, and phenotypic data to fit to something, such as whether a gene is up-regulated or down-regulated in a measurable way. Similarly, on a Web page, one may have text, links to and from, images, html structure, user click behavior, and data about which site the user came from and goes to. A common way to combine these diverse non-numerical data is to put everything into a feature vector. Alternatively, one might construct several similarity graphs or matrices and then combine them in a relatively ad hoc way. The very real danger is that doing so damages the structure in each representation, e.g., linear combinations of the combined data might not be meaningful, or the rule for combining similarity graphs implicitly introduces a lot of descriptive flexibility.

It is thus important to develop principled ways to incorporate metadata on top of a base representation. Such methods should have theoretical foundations, as well as respect the constraints of communication, hardware, and so on.

Relatedly, large-scale data sets may contain information from multiple data sets that were generated individually in very different ways and with different levels of quality or confidence. For example, in a scientific study, multiple experiments may be conducted in different laboratories with varying conclusions. In Web-page analysis, information such as content, click-through information, and link information may be gathered from multiple sources. In multi-media analysis, video, speech, and closed captions are different information sources. In biological applications, if one is interested in a specific protein, information can be obtained from the literature in free-text form describing different research results, or from protein interaction tables, micro-array experiments, the protein's DNA sequence, etc.

One issue is that the most appropriate data representation can differ across these heterogeneous sources. For example, structured data, such as in a relational database, are usually represented in tabular form. Another kind of data, which is becoming increasingly important, is unstructured data such as free-form text and images. These require additional processing before the data can be utilized by computers for statistical inference. Another challenge is how to take advantage of different information sources (or views) to improve prediction or clustering performance. This is referred to as multi-view analysis, a topic that has drawn increasing interest in recent years. Much research has been focused on more sophisticated ways to combine information from different sources instead of naively concatenating them together into a feature vector.

The Challenge of Understanding and Exploiting the Relative Strengths of Data-Oblivious Versus Data-Aware Methods

Among the dimensionality reduction methods, there is a certain dichotomy, with most of the techniques falling into one of two broad categories.

- *Data-oblivious dimensionality reduction* includes methods that compute the dimensionality-reducing mapping without using (or the knowledge of) the data. A prime example of this approach is random projection methods, which select the mapping at random (Johnson and Lindenstrauss, 1984). In this case, the projection is guaranteed to work (in the sense that it preserves the distance structure or other properties) for arbitrary point-sets. In addition, generating such projection requires very little resources in terms of space and/or time, and it can be done before the data are even seen. Finally, this approach leads to results with provable guarantees. Thus, it is popular in areas such as theory of algorithms (Vempala,

2005), sketching algorithms (see Chapter 6), or compressive sensing (Donoho, 2006).

- *Data-aware dimensionality reduction* includes methods that tailor the mapping to a given data set. An example is principal components analysis and its refinements. The mapping is data-dependent in that the algorithm uses the data to compute the mapping, and, as a result, it identifies the underlying structure of the data. Empirically, these methods tend to work better than random projections (e.g., in the sense of preserving more information with shorter sketches), as long as the test data are consistent with the data used to construct the projection. This approach is popular with researchers in machine learning, statistics, and related fields.

A challenge is to merge the benefits of data-oblivious and data-aware dimensionality reduction approaches. For example, perhaps one could develop methods that "typically" achieve the performance of data-aware techniques while also maintaining the worst-case guarantees of random projections (to have something to fall back on in case the data distribution changes rapidly).

The Challenge of Combining Algorithmic and Statistical Perspectives[3]

Given a representation of data, researchers from different areas tend to do very different and sometimes incompatible things with it. For example, a common view of the data in a database, in particular historically among computer scientists interested in data mining and knowledge discovery, has been that the data are an accounting or a record of everything that happened in a particular setting. The database might consist of all the customer transactions over the course of a month, or it might consist of all the friendship links among members of a social networking site. From this perspective, the goal is to tabulate and process the data at hand to find interesting patterns, rules, and associations. An example of an association rule is the (possibly mythical) finding that says that people who buy beer between 5 p.m. and 7 p.m. also buy diapers at the same time. The performance or quality of such a rule is judged by the fraction of the database that satisfies the rule exactly, which then boils down to the problem of finding frequent itemsets. This is a computationally hard problem, and much algorithmic work has been devoted to its exact or approximate solution under different models of data access.

A very different view of the data, more common among statisticians, is that a set of data represents a particular random instantiation of an

---

[3] This section is adapted from Mahoney (2008).

underlying process describing unobserved patterns in the world. In this case, the goal is to extract information about the world from the noisy or uncertain data that is observed. To achieve this, one might posit a model, such as a distribution that represents the variability of the data around its mean. Then, using this model, one would proceed to analyze the data to make inferences about the underlying processes and predictions about future observations. From this perspective, modeling the noise component or variability well is as important as modeling the mean structure well, in large part because understanding the former is necessary for understanding the quality of predictions made. With this approach, one can even make predictions about events that have yet to be observed. For example, one can assign a probability to the event that a given user at a given Web site will click on a given advertisement presented at a given time of the day, even if this particular event does not exist in the database. See Chapter 7 for further discussion of the model-based perspective.

The two perspectives need not be incompatible. For example, statistical and probabilistic ideas are central to much of the recent work on developing improved randomized approximation algorithms for matrix problems. Otherwise-intractable optimization problems on graphs and networks yield to approximation algorithms when assumptions are made about the network participants. Much recent work in machine learning also draws on ideas from both areas. Thus, a major challenge in large-scale data representation is to combine and exploit the complementary strengths of these two approaches. The underlying representation should be able to support in an efficient manner operations required by both approaches. One promising direction takes advantage of the fact that the mathematical structures that provide worst-case guarantees often have fortuitous side effects that lead to good statistical properties (Mahoney and Orecchia, 2011).

REFERENCES

Agarwal, P.K., S. Har-Peled, and K.R. Varadarajan. 2005. Geometric approximation via coresets. *Combinatorial and Computational Geometry.* Volume 52. MSRI Publications, Cambridge University Press, New York, N.Y.

Ailon, N., and B. Chazelle. 2010. Faster dimension reduction. *Communications of the ACM* 53:97-104.

Ailon, N., and E. Liberty. 2011. An almost optimal unrestricted fast Johnson-Lindenstrauss transform. Pp. 185-191 in *Proceedings of the Twenty-Second Annual ACM-SIAM Symposium on Discrete Algorithms.* Society for Industrial and Applied Mathematics, Philadelphia, Pa.

Andoni, A., and P. Indyk. 2008. Near-optimal hashing algorithms for approximate nearest neighbor in high dimensions. *Communications of the ACM* 51(1):117-122.

Asanovic, K., R. Bodik, B.C. Catanzaro, J.J. Gebis, P. Husbands, K. Keutzer, D.A. Patterson, W.L. Plishker, J. Shalf, S.W. Williams, and K.A. Yelick. 2006. *The Landscape of Parallel Computing Research: A View from Berkeley.* University of California, Berkeley, Technical Report No. UCB/EECS-2006-183. December 18. Available at http://www.eecs.berkeley.edu/Pubs/TechRpts/2006/EECS-2006-183.html.

Baeza-Yates, R., and B. Ribeiro-Neto. 1999. *Modern Information Retrieval.* Addison-Wesley Longman, Reading, Mass.

Belkin, M., and P. Niyogi. 2003. Laplacian eigenmaps for dimensionality reduction and data representation. *Neural Computation* 15(6):1373-1396.

Blei, D.M., A.Y. Ng, and M. Jordan. 2003. Latent Dirichlet allocation. *Journal of Machine Learning Research* 3(4-5): 993-1022.

Broder, A. 1997. On the resemblance and containment of documents. P. 21 in *Proceedings of the Compression and Complexity of Sequences 1997.* IEEE Computer Society, Washington, D.C.

Broder, A., and M. Mitzenmacher. 2002. Network applications of Bloom filters: A survey. *Internet Mathematics* 1(4):485-509.

Charikar, M., K. Chen, and M. Farach-Colton. 2002. Finding frequent items in data streams. Pp. 693-703 in *Proceedings of the 29th International Colloquium on Automata, Languages and Programming.* Springer-Verlag, London.

Chung, F. 1992. *Spectral Graph Theory.* American Mathematical Society, Providence, R.I.

Clauset, A., C.R. Shalizi, and M.E.J. Newman. 2009. Power-law distributions in empirical data. *SIAM Review* 51:661-703.

Coifman, R.R., and S. Lafon. 2006. Diffusion maps. *Applied and Computational Harmonic Analysis* 21(July):5-30.

Cormen, T.H., C.E. Leiserson, R.L. Rivest, and C. Stein. 2009. *Introduction to Algorithms,* 3rd edition. MIT Press, Cambridge, Mass.

Cormode, G., and S. Muthukrishnan. 2005. An improved data stream summary: The count-min sketch and its applications. *Journal of Algorithms* 55(1):58-75.

Donoho, D.L. 2006. Compressed sensing. *IEEE Transactions on Information Theory* 52: 1289-1306.

DuMouchel, W. 2002. Data squashing: Constructing summary data sets. Pp. 579-591 in *Handbook of Massive Data Sets* (J. Abello, P.M. Pardalos, and M.G.C. Resende, eds.). Kluwer Academic Publishers, Norwell, Mass.

Ferragina, P., and G. Manzini. 2000. Opportunistic data structures with applications. Pp. 390-398 in *Proceedings of the 41st Annual Symposium on Foundations of Computer Science.* IEEE Computer Society, Washington, D.C.

Garcia-Molina, H., J. Ullman, and J. Widom. 2008. *Database Systems: The Complete Book,* 2nd edition. Prentice-Hall, Upper Saddle River, N.J.

Gersho, A., and R.M. Gray. 1992. *Vector Quantization and Signal Compression.* Springer.

Hansen, P.C. 1987. The truncated SVD as a method for regularization. *BIT* 27(4):534-553.

Hofmann, T., B. Scholkopf, and A.J. Smola. 2008. Kernel methods in machine learning. *Annals of Statistics* 36(3):1171-1220.

Jain, A.K., M.N. Murty, and P.J. Flynn. 1999. Data clustering: A review. *ACM Computing Surveys* 31(3):264-323.

Johnson, W., and J. Lindenstrauss. 1984. Extensions of Lipschitz mappings into a Hilbert space. *Contemporary Mathematics* 26:189-206.

Jones, N., and P. Pevzner. 2004. *An Introduction to Bioinformatics Algorithms.* MIT Press, Cambridge, Mass.

Kumar, S., M. Mohri, and A. Talwalkar. 2009. Sampling techniques for the Nystrom method. Pp. 304-311 in *Proceedings of the Twelfth International Conference on Artificial Intelligence and Statistics* (AISTATS). Available at http://jmlr.org/proceedings/papers/v5/.

Langmead, B., C. Trapnell, M. Pop, and S.L Salzberg. 2009. Ultrafast and memory-efficient alignment of short DNA sequences to the human genome. *Genome Biology* 10(3):R25.

Leskovec, J., K.J. Lang, A. Dasgupta, and M.W. Mahoney. 2009. Community structure in large networks: Natural cluster sizes and the absence of large well-defined clusters. *Internet Mathematics* 6(1):29-123.

Mahoney, M.W. 2008. Algorithmic and statistical challenges in modern large-scale data analysis are the focus of MMDS 2008. *ACM SIGKDD Explorations Newsletter*, December 20.

Mahoney, M.W., and P. Drineas. 2009. CUR matrix decompositions for improved data analysis. *Proceedings of the National Academy of Sciences U.S.A.* 106:697-702.

Mahoney, M.W., and L. Orecchia. 2011. Implementing regularization implicitly via approximate eigenvector computation. Pp. 121-128 in *Proceedings of the 28th International Conference on Machine Learning (ICML)*. ICML, Bellevue, Wash.

Paschou, P., E. Ziv, E.G. Burchard, S. Choudhry, W. Rodriguez-Cintron, M.W. Mahoney, and P. Drineas. 2007. PCA-correlated SNPs for structure identification in worldwide human populations. *PLoS Genetics* 3:1672-1686.

Roweis, S., and L. Saul. 2000. Nonlinear dimensionality reduction by locally linear embedding. *Science* 290(5500):2323-2326.

Salomon, D. 1998. *Data Compression: The Complete Reference*. Springer Verlag, London, U.K.

Spielman, D.A. 2010. Algorithms, graph theory, and linear equations in Laplacian matrices. Pp. 24-26 in *Proceedings of the 39th International Congress of Mathematicians*, Part II. Springer-Verlag, Berlin.

Tenenbaum, J.B., V. de Silva, and J.C. Langford. 2000. A global geometric framework for nonlinear dimensionality reduction. *Science* 290(5500):2319-2323.

Tibshirani, R. 1996. Regression shrinkage and selection via the lasso. *Journal of the Royal Statistical Society: Series B* 58(1):267-288.

van der Maaten, L.J.P., E.O. Postma, and H.J. van den Herik. 2009. *Dimensionality Reduction: A Comparative Review*. Technical Report TiCC-TR 2009-005. Tilburg University, The Netherlands.

Vempala, S.S. 2005. *The Random Projection Method*. DIMACS Series in Discrete Mathematics and Theoretical Computer Science, Volume 65. AMS.

6

Resources, Trade-offs, and Limitations

INTRODUCTION

This chapter discusses the current state of the art and gaps in fundamental understanding of computation over massive data sets. The committee focuses on general principles and guidelines regarding which problems can or cannot be solved using given resources. Some of the issues addressed here are also discussed in Chapters 3 and 5 with a more practical focus; here the focus is on theoretical issues.

Massive data computation uses many types of resources. At a high level, they can be partitioned into the following categories:

- *Computational resources,* such as space, time, number of processing units, and the amount of communication between them;
- *Statistical or information-theoretic resources,* such as the number of data samples and their type (e.g., whether a sample is random or carefully selected by the algorithms, whether the data are "labeled" or "unlabeled," and so on); often, one might also like to minimize the amount and type of information revealed about the data set in order to perform certain computations, to minimize the loss of privacy; and
- *Physical resources,* such as the amount of energy used during the computation.

The use of these resources has been studied in several fields. Reviewing the state of the art in those areas, even only in the context of massive data processing, is a task beyond the scope of this report. Nevertheless, identifying the gaps in current knowledge requires at least a brief review of the background. To this end, the following section begins with a short overview of what theoretical computer science can reveal about the computational resources needed for massive data computations. This is a complement to the background material on statistics in the next two chapters.

RELEVANT ASPECTS OF THEORETICAL COMPUTER SCIENCE

Theoretical computer science studies the strengths and limitations of computational models and processes. Its dual goals are to (1) discover and analyze algorithms for key computational problems that are efficient in terms of resources used and (2) understand the inherent limitations of computation with bounded resources. The two goals are naturally intertwined; in particular, understanding the limitations often suggests approaches for sidestepping them.

In this section the committee surveys an illustrative list of concepts and notions developed in theoretical computer science, with the emphasis on material relevant to computing over massive data sets.[1]

Tractability and Intractability

One of the central notions in understanding computation is that of polynomial time. A problem is solvable in polynomial time if there is an algorithm that can solve it that runs in time N^c for some constant c (i.e., it is $O(N^c)$) on inputs of size N. Polynomial-time algorithms have the following attractive property: doubling the input size results in running time that is only a fixed factor larger (its value depends on c). Therefore, they scale gracefully as the input size increases. This should be contrasted with algorithms with running time exponential in N—e.g., of $O(2^N)$. Here, doubling the input size can increase the running time in a much more dramatic fashion. As such, problems that have polynomial-time algorithms are often referred to as *tractable*. In contrast, problems for which such algorithms are conjectured to not exist are called *intractable*.

Perhaps surprisingly, most of the natural problems are known to fall into one of these two categories. That is, either a polynomial-time algorithm for a problem is known, or the best-known algorithm has exponential running time. Many of the latter problems have the property that if one is given a solution to the problem, it only takes polynomial time to verify whether it is correct. Such problems are said to run in non-deterministic

---

[1] For a broader overview, the Theory Matters blog entry "Vision Nuggets," available at http://thmatters.wordpress.com/vision-nuggets/, provides a good background.

polynomial time (NP). Some such problems have the "one-for-all and all-for-one" property: if any of them can be solved in polynomial time, then all of them can. Those problems are called *NP-complete*. Examples of NP-complete problems include the Traveling Salesman Problem (given a set of cities and distances between them, is there a tour of a given length that visits all cities?) and the Satisfiability Problem (given a set of m constraints over n variables, is there a way to satisfy them all?). The (conjectured) difficulty of such problems comes from the (apparent) need to enumerate an exponential number of possible solutions in order to find the feasible one.

Although it is not known whether an NP-complete problem can be solved in polynomial time—this question, called "P versus NP," is one of the central open problems in computer science—it is widely conjectured that such algorithms do not exist.[2] The notion of NP-completeness thus provides a very useful tool guiding algorithm design. Specifically, showing that a problem is NP-complete means that instead of trying to find a complete solution, one likely needs to modify the question. This can be done, for example, by allowing approximate answers or exploiting the particular structure of inputs. For a recent overview of such developments in the context of satisfiability, see Malik and Zhang (2009).

Given the usefulness of NP-completeness and other computational hardness tools when dealing with computational problems, it is natural to explore their uses in the context of massive data sets. This question is examined in more depth later in this chapter.

Sublinear, Sketching, and Streaming Algorithms

In the search for efficient algorithms for large-scale problems, researchers formulate more stringent models of computation. One such notion that is particularly relevant to massive data is that of sublinear algorithms. They are characterized as using an amount of resources (e.g., time or space) that is much smaller than the input size, often exponentially smaller. This, in particular, means that the algorithm cannot read or store the whole input, and instead it must extrapolate the answer from the small amount of information read or stored.

One of the popular computational models of this type is data-stream computing. In the data-stream model, the data need to be processed "on the fly"—i.e., the algorithm can make only a single pass over the data, and the storage used by the algorithm can be much smaller than the input size. Typically, streaming algorithms proceed by computing a summary or "sketch" of the input, which is much shorter but nevertheless sufficient to approximate the desired quantity. Perhaps surprisingly, for many problems,

---

[2] Fortnow (2009) provides an overview.

efficient sketching methods are known to exist (Muthukrishnan, 2005; Indyk, 2007). For example, consider the problem of counting the number of distinct elements in a stream. This task is known to require space that is at least as large as the actual number of distinct items; the items have to be stored temporarily to avoid double-counting the items already seen. However, it is possible to approximate this quantity using only a logarithmic amount of space (Flajolet and Martin, 1985).

Other models of sublinear computation include sublinear time computation (where the algorithm is restricted to using an amount of time that scales sublinearly with the input size) and approximate property testing (where one tests whether the input satisfies a certain property using few data samples). See Czumaj and Sohler (2010) or Rubinfeld and Shapira (2011) for an overview.

Communication Complexity

The field of communication complexity (Kushilevitz and Nisan, 1997) studies the amount of information that needs to be extracted from the input, or communicated between two or more parties sharing parts of the input, to accomplish a given task. The aforementioned sketching approach to sublinear computation is one of the studied models, but many other models have been investigated as well. In contrast to NP-completeness, communication complexity techniques make it possible to prove that some tasks cannot be accomplished using limited communication. For example, consider the following set disjointness problem, where two parties want to determine whether two data sets of equal size, each held locally by one of the parties, contain any common items. It has been shown that in order to accomplish this task, the parties must exchange a number of bits that is linear in the size of the data set (Razborov, 1992; Kalyanasundaram and Schnitger, 1992).

External Memory

Another way of modeling space-limited computation is to focus on the cost of transferring data between the fast local memory and slow external memory (e.g., a disk). This approach is motivated by the fact that, in many scenarios, the transfer cost dominates the overall running time. The external memory model (Vitter, 2008) addresses precisely that phenomenon. Specifically, the computer system is assumed to be equipped with a limited amount of main memory (which is used to perform the computation) and an unbounded external memory such as disk drive (which stores the input and any intermediate data produced by the algorithm). The data are exchanged between the main and external memories via a sequence of input/output

(I/O) operations. Each such operation transfers a contiguous block of data between the memories. The complexity of an algorithm is then measured by the total number of I/O operations that the algorithm performs.

The algorithms that are efficient in the external memory model minimize the need to refer to data that are far apart in memory storage or in time. This approach enables the computation to limit the required number of block transfers. See Vitter (2008) for an overview.

In general, external memory algorithms are "cache-aware"; that is, they must be supplied with the amount of available main memory before they can proceed. This drawback is removed by "cache-oblivious algorithms" (Frigo et al., 1999), which automatically adapt to the amount of memory (in fact, general caching mechanism) available to the algorithm.

Parallel Algorithms

A fundamental and widely studied question is to understand for which problems one can obtain a speedup using parallelism. Many models of parallel computation have been studied. Perhaps the one that has attracted the greatest amount of attention is the class of problems having polynomial-time sequential algorithms for which one can obtain exponential speedups by using parallelism. Such speedups are known to be possible for surprisingly many problems, such as finding a perfect matching in a graph. That problem calls for finding a subset of edges that contains exactly one edge incident to any vertex, given a set of nodes and edges between them. There are also problems that are conjectured to be inherently sequential, i.e., they appear to not be amenable to exponential speedups.

Computational Learning Theory

The field of computational learning theory (Blum, 2003) studies the computational aspects of extracting knowledge from data. Specifically, it addresses the following question: How much data and computational resources are needed in order to "learn" a concept of interest with a given accuracy and confidence? For example, a well-studied task is to infer a linear classifier that can separate data into positive and negative classes, given a sequence of labeled examples and using a bounded amount of computational resources. Variations of the basic framework include semi-supervised learning (where the labels are specified for only some of the examples) and active learning (where the algorithm can query the value of a label for some or all examples). Computational learning theory has natural connections to statistics, especially statistical learning theory, and utilizes and builds on notions from those fields.

GAPS AND OPPORTUNITIES

Despite the extensive amount of work devoted to the topic, the fundamentals of computation over massive data sets are not yet fully understood. This section examines some of the gaps in the current knowledge and possible avenues of addressing them.

Challenges for Computer Science

Computational Hardness of Massive Data Set Problems

Given the usefulness of computational hardness in guiding the development of general polynomial-time algorithms, it would be helpful to be able to apply such tools to algorithms designed for massive data as well. However, polynomial-time is typically not a sufficient condition for tractability when the input to a problem is very large. For example, although an algorithm with running time N^4 can be quite efficient for moderate values of N (say, a few thousand), this is no longer the case when N is of the order of billions or trillions. One must therefore refine the approach. This will involve (1) defining more refined boundaries between the tractable and the intractable that model the massive-data computation more accurately and (2) identifying new "hard" problems that are (conjectured to be) unsolvable within those boundaries.

The class of sublinear algorithms presented in the earlier section is a well-studied example of this line of research. However, the limitations imposed by that class are quite restrictive, and they exclude some tasks (such as sorting) that can be successfully performed even on very large data sets. Two examples of more expressive classes of problems are those requiring sub-quadratic time and those requiring linear time.

Quadratic time is a natural boundary of intractability for problems over massive data because many problems have simple quadratic-time solutions. For example, the generalized N-body problem class discussed in Chapter 10—which consists of problems involving interactions between pairs of elements—and the class of alignment problems also discussed in that chapter are amenable to such algorithms. For massive data, however, one typically needs algorithms that run in time faster than quadratic. In some cases, obtaining better algorithms is possible. For example, the basic N-body problem (involving particles interacting in a three-dimensional space) can be solved in $O(N \log N)$ time by use of the Fast Multipole Method.

Unfortunately, unlike for polynomial-time computation, versatile tools are lacking that would help determine whether a given problem has a sub-quadratic-time algorithm or not. A few proposals for such tools exist in the literature. For example, the 3SUM problem (given a set of N numbers, are there three numbers in the set that sum to zero?) is a task that appears to be unsolvable in less than quadratic time (Gajentaan and Overmars, 1995). As a result, 3SUM plays a role akin to that of an NP-complete problem: if 3SUM can efficiently be reduced to a problem of interest, this indicates that the problem cannot be solved in less than quadratic time. Several such reductions are known, especially for computational-geometry and pattern-matching problems. However, the "web of reductions" is still quite sparse, especially when compared to the vast body of work on polynomial time/NP-complete problems. A better understanding of such reductions is sorely needed.

Linear running time is a gold standard of algorithmic efficiency. As long as an algorithm must read the whole input to compute the answer, it must run in at least linear time. However, as in the case of sub-quadratic-time algorithms, there are no methods that would indicate whether a problem is likely to have a linear time solution or not.

One problem that has been identified as "hard" in this regime is the problem of computing the discrete Fourier transform. The Fast Fourier Transform algorithm performs this task in time $O(N \log N)$, and despite decades of research no better algorithm is known. Thus, if a given problem requires computation of the discrete Fourier transform, that is a strong indication that (at the very least) it will be difficult to obtain a linear time algorithm for that problem. Still, it would be helpful to develop a better foundation for the study of such problems, for example, by developing a richer or more refined set of problems that are conjectured to be hard.

More study is needed for each of these classes of algorithms.

The Role of Constants

To this point, the discussion of running times involved asymptotic analysis—that is, the running times were specified up to a leading constant. Even though it is generally understood that the value of the leading constants can make a difference between a practical algorithm and an unfeasible "galactic algorithm" (Lipton, 2010), asymptotic analysis nevertheless remains the standard theoretical tool. Optimizing constant factors is often thought to belong to algorithm engineering rather than algorithm design. A notable exception to this trend includes the study of classic problems like median finding or sorting, especially in the context of average-case analysis.

There are several good reasons for this state of affairs. For one, the asymptotic running time provides a simple and convenient way to describe, compare, and reason about the algorithm performance: one has to deal with only one parameter, namely the exponent. Moreover, the actual running times of an algorithm can be highly variable, even for a particular input, because they are both time-dependent (computers gain processing power every few months) and platform-dependent (different instructions can have different execution times on different machines). All of these reasons motivate a constant-free approach to running time analysis, unless the cost model is very well defined.

At the same time, there exist potential opportunities in trying to understand constants in more depth. For one, ignoring constant factors can obscure the dependencies on implicit data parameters, such as the dimension of the underlying space, precision, etc. Moreover, some of the aforementioned motivating factors are no longer as relevant. For example, it is no longer the case that computers are getting faster: the increase in processing power in coming years is projected to come from increased parallelism rather than clock speed.

The platform-dependence issue continues to be significant. However, as described in Chapter 10, it is often the case that an algorithm is assembled from a relatively small set of building blocks, as opposed to being designed entirely from scratch. In such scenarios it could be possible to encapsulate platform dependence in implementations of those blocks, while making the rest of the algorithm platform-independent. The number of times the subroutine is invoked could then be optimized in a platform-independent fashion.

New Models for Massive Data Computation

Understanding the quickly evolving frameworks and architectures for massive data processing will likely require constructing and investigating new models of computation. Many such models have been designed in recent years, especially in the context of parallel data processing (see Chapter 3). This includes models such as MapReduce, Hadoop and variations, multicores, graphic processing units (GPUs), and parallel databases. The new models and their relationship to more traditional models of parallel computation have already been a subject of theoretical studies.[3] However, more work is certainly needed.

Challenges for Other Disciplines

Perhaps the biggest challenge to understanding the fundamentals of computing over massive data lies in understanding the trade-offs between resources traditionally studied by computer science and those typically

---

[3] For example, see the website for DIMACS Workshop on Parallelism: A 2020 Vision, March 14-16, 2011, Piscataway, N.J., available at http://dimacs.rutgers.edu/Workshops/Parallel/slides/slides.html.

studied by statistics or physics. This section examines some of the issues that lie at these intersections.

Statistics

Traditionally, computer sciences view the input data as a burden: the larger it is, the more work that needs to be done to process it. If one views the input as a "blob" of arbitrary data, this conclusion appears inevitable. However, if one assumes that the data have some statistical properties—that is, they are a sequence of samples from some distribution or that the data have sparsity or other structural properties—then the result might be different. In fact, some problems have the property that the more data are available, the easier it becomes to solve them.

Quantitative trade-offs of this type have been investigated, for example, in computational learning theory (Blum, 2003), machine learning (Bottou and Bousquet, 2008), and sublinear algorithms (Chien et al., 2010; Harsha et al., 2004). However, many questions remain open. For example, in computational learning theory, the computational limitations are typically specified in terms of polynomial-time computability, and thus the limitations of that topic (as discussed earlier) apply.

Another issue that spans statistical and computational aspect of massive data is privacy. This line of research addresses the question of how much information about the data must be revealed in order to perform some computation or answer some queries about the data. The problem has been extensively studied in statistics and, more recently, in applied (Sweeney, 2002) and theoretical (Dwork, 2006) computer science. Privacy is a very large topic in its own right, and this report does not attempt to address the privacy issues associated with massive data and its analysis.

Physical Resources

Computation is ultimately a physical phenomenon, consuming and emitting energy. Over the past decade, this aspect of computation has attracted renewed attention. There are several factors responsible for this state of affairs:

- The large amount of consumed and dissipated energy is the key reason why the steady increase in processor clock rates has slowed in recent years.
- The ubiquity of energy-limited mobile computing devices (smart phones, sensors, and so on) has put a premium on optimizing energy use.

- The impact of computation and data storage on the environment has motivated the development of green computing (Hölzle and Weihl, 2006).

A fair amount of theoretical research has been devoted to reversible computing (Bennett, 1973), which aims to understand the necessary condition for computation to be energy efficient. However, the lower bounds for energy use in computation, and the trade-offs between energy efficiency and computation time, are still not fully understood (Snir, 2011). For example, what are the lower bounds on the amount of energy required to perform basic algorithmic tasks, such as sorting? While there have been a number of applied studies aimed at finding energy efficient architectures and algorithms for sorting,[4] no non-trivial lower bounds for energy consumptions appear to be known. In fact, even formulating this question rigorously presents a challenge, because the input and the output to the problem can take many physical forms. More research is needed to clarify the model, limitations, and trade-offs.

REFERENCES

Bennett, C.H. 1973. Logical reversibility of computation. *IBM Journal of Research and Development* 17(6):525-532.

Blum, A. 2003. "Tutorial on Machine Learning Theory." Presented at the 44th Annual IEEE Symposium on Foundations of Computer Science, October 11-14, Cambridge, Mass. (FOCS 2003). Available at http://www.cs.cmu.edu/~avrim/Talks/FOCS03/index.html.

Bottou, L., and O. Bousquet 2008. The tradeoffs of large scale learning. Pp. 161-168 in *Advances in Neural Information Processing Systems 20* (J.C. Platt, D. Koller, Y. Singer, and S. Roweis, eds.). NIPS 2007 Online Papers. NIPS Foundation. Available at http://books.nips.cc/nips20.html.

Chien, S., K. Ligett, and A. McGregor. 2010. Space-efficient estimation of robust statistics and distribution testing. Pp. 251-265 in *Proceedings of Innovations in Computer Science*. Tsinghua University Press, Tsinghua, China.

Czumaj, A., and C. Sohler. 2010. "Sublinear-time Algorithms." Pp. 42-66 in *Property Testing. Current Research and Surveys* (O. Goldreich, ed.). LNCS 6390. Springer-Verlag, Berlin, Heidelberg.

Dwork, C. 2006. Differential privacy. Pp. 1-12 in *33rd International Colloquium on Automata, Languages and Programming, Part II* (ICALP 2006). Springer Verlag, New York, N.Y.

Flajolet, P., and G. Martin. 1985. Probabilistic counting algorithms for data base applications. *Journal of Computer and System Sciences* 31(2):182-209.

Fortnow, L. 2009. The status of the P versus NP problem. *Communications of the ACM* 52(9):78-86.

---

[4] See, e.g., Sort Benchmark Home Page, available at http://sortbenchmark.org/.

Frigo, M., C.E. Leiserson, H. Prokop, and S. Ramachandran. 1999. Cache-oblivious algo-rithms. Pp. 285-298 in *Proceedings of the 40th Annual Symposium on Foundations of Computer Science*. IEEE Computer Society, Washington, D.C. Available at http://ieee xplore.ieee.org/xpl/tocresult.jsp?isnumber=17631&isYear=1999.

Gajentaan, A., and M. Overmars. 1995. On a class of $O(n^2)$ problems in computational ge-ometry. *Computational Geometry: Theory and Applications* 5(3):165-185.

Harsha, P., Y. Ishai, J. Kilian, K. Nissim, and S. Venkatesh. 2004. Communication versus com-putation. Pp. 745-756 in *Proceedings of the 31st International Colloquium on Automata, Languages and Programming* (ICALP). Springer, New York, N.Y.

Hölzle, U., and B. Weihl. 2006. *High-Efficiency Power Supplies for Home Computers and Servers*. Internal Google technical report. Available at http://static.googleusercontent. com/external_content/untrusted_dlcp/services.google.com/en/us/blog_resources/PSU_ white_paper.pdf.

Indyk, P. 2007. "Sketching, Streaming and Sublinear-Space Algorithms." Graduate course notes. Available at http://stellar.mit.edu/S/course/6/fa07/6.895/.

Kalyanasundaram, B., and G. Schnitger. 1992. The probabilistic communication complexity of set intersection. *SIAM Journal on Discrete Mathematics* 5(5):545-557.

Kushilevitz, E., and N. Nisan. 1997. *Communication Complexity*. Cambridge University Press.

Lipton, R. "Galactic Algorithms." Online posting. Available at http://rjlipton.wordpress. com/2010/10/23/galactic-algorithms/ or http://stellar.mit.edu/S/course/6/fa07/6.895/.

Malik, S., and L. Zhang. 2009. Boolean satisfiability from theoretical hardness to practical success. *Communications of the ACM* 52(8):76-82.

Muthukrishnan, S. 2005. *Data Streams: Algorithms and Applications*. Now Publishers, Hanover, Mass.

Razborov, A.A. 1992. On the distributional complexity of disjointness. *Theoretical Computer Science* 106(2):385-390.

Rubinfeld, R., and A. Shapira. 2012. Sublinear time algorithms. *SIAM Journal of Discrete Mathematics* 25(4):1562-1588.

Snir, M. 2011. "Parallel Computing 2020: Preparing for the Post-Moore Era." Presentation from the DIMACS Workshop on Parallelism: A 2020 Vision. Available at http://dimacs. rutgers.edu/Workshops/Parallel/slides/snir.pdf.

Sweeney, L. 2002. k-anonymity: A model for protecting privacy. *International Journal on Uncertainty, Fuzziness and Knowledge-based Systems* 10(5):557-570.

Vitter, J. 2008. Algorithms and data structures for external memory. *Foundations and Trends in Theoretical Computer Science* 2(4):305-474.

7

Building Models from Massive Data

INTRODUCTION TO STATISTICAL MODELS

The general goal of data analysis is to acquire knowledge from data. Statistical models provide a convenient framework for achieving this. Models make it possible to identify relationships between variables and to understand how variables, working on their own and together, influence an overall system. They also allow one to make predictions and assess their uncertainty.

Statistical models are usually presented as a family of equations (mathematical formulas) that describe how some or all aspects of the data might have been generated. Typically these equations describe (conditional) probability distributions, which can often be separated into a systematic component and a noise component. Both of these components can be specified in terms of some unknown model parameters. These model parameters are typically regarded as unknown, so that they need to be estimated from the data. For a model to be realistic and hence more useful, it will typically be constrained to honor known or assumed properties of the data. For example, some measurements might always be positive or take on values from a discrete set. Good model building entails both specifying a model rich enough to embody structure that might be of use to the analyst and using a parameter estimation technique that can extract this structure while ignoring noise.

Data-analytic models are rarely purely deterministic—they typically include a component that allows for unexplained variation or "noise." This noise is usually specified in terms of *random variables,* that is, variables whose values are not known but are generated from some probability distribution. For example, the number of people visiting a particular website on a given day is random. In order to "model"—or characterize the distribution of—this random variable, statistical quantities (or parameters) might be considered, such as the average number of visits over time, the corresponding variance, and so on. These quantities characterize long-term trends of this random variable and, thus, put constraints on its potential values. A better model for this random variable might take into account other observable quantities such as the day of the week, the month of the year, whether the date is near some major event, and so on. The number of visits to the website can be constrained or predicted by these additional quantities, and their relationship will lead to a better model for the variable. This approach to data modeling can be regarded as statistical modeling: although there are no precise formulas that can deterministically describe the relationship among observed variables, the distribution underlying the data can be characterized. In this approach, one can only guess a certain form of the relationship up to some unknown parameters, and the error— or what is missed in this formulation—will be regarded as noise. Statistical modeling represents a powerful approach for understanding and analyzing data (see McCullagh, 2002).

In what follows, the committee does not make a sharp distinction between "statistics" and "machine learning" and believes that any attempt to do so is becoming increasingly difficult. Statisticians and machine learners work on similar problems, albeit sometimes with a different aesthetic and perhaps different (but overlapping) skill sets. Some modeling activities seem especially statistical (e.g., repeated measures analysis of variance), while others seem to have more of a machine-learning flavor (e.g., support vector machines), yet both statisticians and machine learners can be found at both ends of the spectrum. In this report, terms like "statistical model" or "statistical approach" are understood to *include* rather than *exclude* machine learning.

There are two major lenses through which statistical models are framed, which are described briefly below.

The Frequentist View

The first viewpoint is from classical statistics, where models can take a variety of forms, as can the methods for estimation and inference. One might model the conditional mean of the response (target for prediction) as a parametrized function of the predictors (e.g., linear regression). Although not a requirement, this model can be augmented with an additive noise component to specify the conditional distribution of the response given the predictors. Logistic regression models the conditional distribution of a categorical response given predictors, again using a parametrized model for the conditional probabilities. Most generally, one can specify a joint distribution for both response and predictors. Moving beyond regression, multivariate models also specify a joint distribution, but without distinguishing response and predictor variables.

The frequentist approach views the model parameters as unknown constants and estimates them by matching the model to the available training data using an appropriate metric. Minimizing the empirical sum-of-squared prediction errors is popular for regression models, but other metrics (sum of absolute errors, trimmed sum-of squares, etc.) are also reasonable. Maximum likelihood estimation is a general approach when the model is specified via a (conditional) probability distribution; this approach seeks parameter estimates that maximize the probability of the observed data.

The goal of the statistician is to estimate the parameters as accurately as possible and to make sure the fit of the model to the observed data is satisfactory. The notion of accuracy is based on average frequencies. For example, it should be the case that, on average, over many possible draws of similar data from the model, the estimation procedure yields a value of the parameter that is close to the true underlying parameter. Thus mean-squared error, sampling variance, and bias are used to characterize and rate different estimation procedures. It should also be the case that were the sample size to increase, the estimate should converge to the true parameter, which is defined in terms of the limiting empirical distribution of the data. This is the frequentist concept of consistency.

As an example, consider the problem of modeling customer arrivals at a service center. One can model the data using a Poisson distribution, with an unknown parameter r indicating the daily arrival rate. If one estimates r from the historic observations, then the approximate behavior (probability distribution) of the observations (customer visits per day) is specified. The estimate is random, because it depends on the random data, and one would like for it to be close to the true parameter when averaging over this randomness, that is, when considering all possible draws from the Poisson distribution.

The frequentist view focuses on the analysis of estimation procedures and is somewhat agnostic about the nature of the procedures that are considered. That said, the analysis often suggests the form of good procedures (even optimal procedures). For example, members of a broad class of estimation procedures, known as M-estimators, take the form of optimization problems. This class includes maximum likelihood estimation, but is much broader.

The Bayesian View

The second viewpoint is from Bayesian statistics. In this case, the model specifies a family of (conditional) probability distributions, indexed by parameters. These parameters are considered random as well, and so a prior distribution needs to be specified for them. For website visits, for example, it might be assumed that the rate parameter r has a flat distribution over the positive interval $(0,\infty)$, or that it decays as $1/r$. Once a prior is assumed, the joint distribution over the parameters and the observations is well defined. The distribution of the model parameter is characterized by the *posterior distribution* of the parameters given historic data, which is defined as the conditional distribution of the parameter given the data. The posterior can be calculated from Bayes's theorem. Bayesian models replace the "parameter estimation" problem by the problem of defining a "good prior" plus computation of the posterior, although when coupled with a suitable loss function, Bayesian approaches can also produce parameter estimates. The concept of a loss function is discussed in later sections.

From a procedural point of view, Bayesian methods often take the form of numerical integration. This is because the posterior distribution is a normalized probability measure, and computing the normalization factor requires integration. The numerical integration is generally carried out via some form of Monte Carlo sampling procedure or other type of numerical approximation.

Thus, procedurally, methods studied in frequentist statistics often take the form of optimization procedures, and methods studied in Bayesian statistics often take the form of integration procedures. The relevance of these viewpoints to massive data analysis comes down in part to the scalability of optimization versus integration. (That said, it is also possible to treat integration problems using the tools of optimization; this is the perspective of the variational approach to Bayesian inference (Wainwright and Jordan, 2003).)

It should also be noted that there are many links between the frequentist and Bayesian views at a conceptual level, most notably within the general framework of statistical decision theory. Many data analyses involve a blend of these perspectives, either procedurally or in terms of the analysis.

Nonparametrics

Although the committee has framed its discussion of models in terms of "parameters," one often sees a distinction made between "parametric models" and "nonparametric models," perhaps confusingly, because both classes of models generally involve parameters. The difference is that in a parametric model, the number of parameters is fixed once and for all, irrespective of the number of data points. A nonparametric model is (by definition) "not parametric," in that the number of parameters is not fixed, but rather grows as a function of the number of data points. Sometimes the growth is explicitly specified in the model, and sometimes it is implicit. For example, a nonparametric model may involve a set of functions that satisfy various constraints (e.g., smoothness constraints), and the choice of parameters (e.g., coefficients in a Fourier expansion) may be left implicit. Moreover, the growth in the number of parameters may arise implicitly as part of the estimation procedure.

Although parametric models will always have an important role to play in data analysis, particularly in situations in which the model is specified in part from an underlying scientific theory (such that the parameters have meaning in the theory), the committee finds that the nonparametric perspective is particularly well aligned with many of the goals of massive data analysis. The nonparametric perspective copes naturally with the fact that new phenomena often emerge as data sets increase in size.

The distinction between parametric and nonparametric models is orthogonal to the frequentist/Bayesian distinction—there are frequentist approaches to nonparametric modeling and Bayesian approaches to nonparametric modeling. The Bayesian approach to nonparametrics generally involves replacing classical prior distributions with stochastic processes, thereby supplying the model with an open-ended (infinite) number of random parameters. The frequentist approach, with its focus on analysis, shoulders the burden of showing that good estimates of parameters can be obtained even when the number of parameters is growing.

It is also possible to blend the parametric and nonparametric perspective in a hybrid class of models referred to as "semiparametric." Here the model may be based on two subsets of parameters, one which is fixed and which is often of particular interest to the human data analyst, and the other which grows and copes with the increasing complexity of the data as a function of the size of the data set.

Loss Functions and Partially Specified Models

One rarely wishes to model all aspects of a data set in detail, particularly in the setting of massive data. Rather, there will aspects of the model that more important than others. For example, particular subsets of the parameters may be of great interest and others of little interest (the latter are often referred to as "nuisance parameters"). Also, certain functions of parameters may be of particular interest, such as the output label in a classification problem. Such notions of importance or focus can be captured via the notion of a "loss function." For example, in binary classification, one may measure performance by comparing the output of a procedure (a zero or one) with the true label (a zero or one). If these values disagree, then the loss is one, otherwise it is zero. In regression it is common to measure the error in a fitted function via the squared-error loss.

Both frequentists and Bayesians make use of loss functions. For frequentists the loss is a fundamental ingredient in the evaluation of statistical procedures; one wishes to obtain small average loss over multiple draws of a data set. The probabilistic model of the data is used to define the distribution under which this average is taken. For Bayesians the loss function is used to specify the aspects of the posterior distribution that are of particular interest to the data analyst and thereby guide the design of posterior inference and decision-making procedures.

The use of loss functions encourages the development of partially specified models. For example, in regression, where the goal is to predict Y from X, if the loss function only refers to Y, as is the case with the least-squares loss, then this encourages one to develop a model in which the distribution of X is left unspecified. Similarly, in binary classification, where the goal is to label a vector X with a zero or one, one can forgo an attempt to model the class-conditional distributions of X and focus only on a separating surface that predicts the labels well. For example, if one considers the class of all hyperplanes, then one has a parametric model in which the parameters are those needed to define a hyperplane (note that the number of parameters is fixed in advance in this case). Alternatively, one can consider flexible surfaces that grow in complexity as data accrue; this places one in the nonparametric modeling framework. Note that the use of the term "model" has become somewhat abstract at this point; the parameters have a geometric interpretation but not necessarily a clear interpretation in the problem domain. On the other hand, the model has also become more concrete in that it is targeted to the inferential goal of the data analyst via the use of a loss function.

Even if one makes use of a partially specified model in analyzing the data, one may also have in mind a more fully specified probabilistic model for evaluating the data analysis procedure; that is, for computing the "average" loss under that procedure. While classically the model being estimated and the model used for evaluation were the same, the trend has been to separate these two notions of "model." A virtue of this separation is that it makes it possible to evaluate a procedure on a wider range of situations than the situation for which it was nominally designed; this tests the "robustness" of the procedure.

Other Approaches

In addition to the statistical modeling perspectives discussed above, which have been extensively studied in the statistical and machine-learning literature, other forms of data-analysis algorithms are sometimes used in practice. These methods also give meaningful descriptions of the data, but they are more procedure-driven than model-driven.

Some of these procedures rely on optimization criteria that are not based on statistical models or even have any underlying statistical underpinning. For example, the k-means algorithm is a popular method for clustering, but it is not based on a statistical model of the data. However, the optimization criterion still characterizes what a data analyst wants to infer from the data: whether the data can be clustered into coherent groups. This means that instead of a statistical model, appropriately defined optimization formulations may be more generally regarded as models that capture useful descriptions of the data. In such a case, parameters in the optimization formula determine a model, and the optimal parameter gives the desired description of the data. It should be noted that many statistical parameter estimation methods can be regarded as optimization procedures (such as maximum-likelihood estimation). Therefore, there is a strong relationship between the optimization approach (which is heavily used in machine learning) and the more traditional statistical models.

Some other data-analysis procedures try to find meaningful characterizations of the data that satisfy some descriptions but are not necessarily based on optimization. These methods may be considered as algorithmic approaches rather than models. For example, methods for finding the most frequent items in a large-scale database (the "heavy hitters") or highly correlated pairs are computational in nature. Although the specific statistical quantities these algorithms try to compute provide models of the data in a loose sense (i.e., as constraints on the data), the focus of these methods is on computational efficiency, not modeling. Nevertheless, the algorithmic approaches are important in massive data analysis simply because the need for computational efficiency goes hand-in-hand with massive data analysis.

DATA CLEANING

In building a statistical model from any data source, one must often deal with the fact that data are imperfect. Real-world data are corrupted with noise. Such noise can be either systematic (i.e., having a bias) or random (stochastic). Measurement processes are inherently noisy, data can be recorded with error, and parts of the data may be missing. The data produced from simulations or agent-based models, no matter how complex, are also imperfect, given that they are built from intuition and initial data. Data can also be contaminated or biased by malicious agents. The ability to detect false data is extremely weak, and just having a massive quantity of data is no guarantee against deliberate biasing. Even good data obtained by high-quality instrumentation or from well-designed sampling plans or simulations can produce poor models in some situations. Noisy and biased data are thus unavoidable in all model building, and this can lead to poor predictions and to models that mislead.

Although random noise can be averaged out, loosely speaking, using multiple independent experiments—that is, the averaged noise effect approaches zero—this is not the case with systemic noise. In practice, both kinds of noise exist, and thus both kinds should be incorporated into models of the data. The goal is often to build statistical models that include one or more components of noise so that noise can be separated from signal, and thus relatively complex models can be used for the signal, while avoiding overly complex models that would find structure where there is none. The modeling of the noise component impacts not only the parameter estimation procedure, but also the often informal process of cleaning the data and assessing whether the data are of high-enough quality to be used for the task at hand. This section focuses on this informal activity of data cleaning in the setting of massive data.

The science and art of cleaning data is fairly well developed when data sets are of modest size,[1] but new challenges arise when dealing with massive data. In small-scale applications, data cleaning often begins with simple sanity checking. Are there any obvious mistakes, omissions, mislabelings, and so on, that can be seen by sampling a small subset of the data? Do any variables have obviously incorrect values? This kind of checking typically involves plotting the data in various ways, scanning through summaries, and producing particular snapshots that are designed to expose bad data. Often the result of this process is to return to the source to confirm suspicious entries, or to fill in omitted fields.

How does this approach change with massive data? The sanity checking and identification of potential problems can still be performed using samples and snapshots, although determining how to find representative samples can sometimes pose problems (see Chapter 8). However, the ability to react to issues will be constrained by time and size, and most human intervention is impossible. There are at least two general approaches to overcoming this problem:

- One can build auto-cleaning mechanisms into data-capture and data-storage software. This process requires monitoring via reporting and process-control mechanisms. Ideally, an audit trail should accompany the data, so that changes can be examined and reversed where needed.
- Certain problems can be anticipated and models built that are resistant to those problems.

---

[1] See, e.g., Dasu and Johnson (2003).

Although there has been a lot of research in "robust" modeling in accordance with the second approach, the vast majority of models currently in use are not of this kind, and those that are tend to be more cumbersome. Hence the first approach is more attractive, but the second should be considered.

For example, features that are text-based lead to many synonyms or similar phrases for the same concept. Humans can curate these lists to reduce the number of concepts, but with massive data this probably needs to be done automatically. Such a step will likely involve natural language processing methodology, and it must be sufficiently robust to handle all cases reasonably well. With storage limitations, it may not be feasible to store all variables, so the method might have to be limited to a more valuable subset of them. This subset may well be the "cleanest." With text-based features, some are so sparse they are ultimately useless and can be cleaned out of the system.

Missing data is often an issue, and dealing with it can be viewed as a form of cleaning. Some statistical modeling procedures—such as trees, random forests, and boosted trees—have built-in methods for dealing with missing values. However, many model-building approaches assume the data are complete, and so one is left to impute the missing data prior to modeling. There are many different approaches to data imputation. Some simple methods that are practical on a massive scale are to replace the missing entries by the mean for that variable. This implicitly assumes that the omissions are completely random. Other more sophisticated methods treat the missing data imputation as a prediction problem—predicting the missing entries for a variable using that variable as a response, and all the values as input variables. This might create a computational burden that is prohibitive with massive data, so good compromises are sought.

Whatever approaches are used to clean and preprocess the data, the steps should be documented, and ideally the scripts or code that were used should accompany the data. Following these steps results in a process that is reproducible and self-explanatory.

CLASSES OF MODELS

Data analysts build models for two basic reasons: to understand the past and to predict the future. One would like to understand how the data were generated, the relationships between variables, and any special structure that may exist in the data. The process of creating this understanding is often referred to as unsupervised learning. A more focused task is to build a prediction model, which allows one to predict the future value of a target variable as a function of the other variables at one's disposal, and/or at a future time. This is often referred to as *supervised learning*.

Data-generating mechanisms typically defy simple characterization, and thus models rarely capture reality perfectly. However, the general hope is that carefully crafted models can capture enough detail to provide useful insights into the data-generating mechanism and produce valuable predictions. This is the spirit behind the famous observation from the late George Box that all models are wrong, but some are useful. While the model-building literature presents a vast array of approaches and spans many disciplines, model building with massive data is relatively uncharted territory. For example, most complex models are computationally intensive, and algorithms that work perfectly well with megabytes of data may become infeasible with terabytes or petabytes of data, regardless of the computational power that is available. Thus, in analyzing massive data, one must re-think the trade-offs between complexity and computational efficiency.

This section provides a summary of major techniques that have been used for data mining, statistical analysis, and machine learning in the context of large-scale data, but which need re-evaluation in the context of massive data. Some amplification is provided in Chapter 10, which discusses computational kernels for the techniques identified here.

Unsupervised Learning

Unsupervised learning or data analysis aims to find patterns and structures in the data. Some standard tasks that data analysts address include the following:

- Clustering, which is partitioning data into groups so that data items within each group are similar to each other and items across different groups are not similar. K-means and hierarchical clustering are popular algorithmic approaches. Mixture models use a probabilistic framework to model clusters.
- Dimension reduction, which finds a low-dimensional space that approximately contains the data; another view is to represent high-dimensional data points by points in a lower-dimensional space so that some properties of the data can be preserved. For example, one approach might be to preserve enough information to fully reconstruct the data, and another may be to preserve only enough information to recover distances among data points. Dimension reduction can be either linear or nonlinear depending on the underlying model. Statistically, dimension reduction is closely related to factor analysis. Factor models treat dimensions as factors, and each observation is represented by a combination of these factors.
- Anomaly detection, or determining whether a data point is an outlier (e.g., is very different from other typical data points). This

is often achieved by defining a criterion that characterizes how a typical data point in the data set behaves; this criterion is then used to screen all the points and to flag outliers. One general approach is to use a statistical model to characterize the data, and an outlier is then a point that belongs to a set with a small probability (which can be measured by a properly defined p-value) under the model.

- Characterizing the data through basic statistics, such as mean, variance, or high-order moments of a variable, correlations between pairs of variables, or the frequency distribution of node degrees in a graph. Although simple from a modeling perspective, the main challenge of these methods is to find computational algorithms that can efficiently work with massive data.
- Testing whether a probability model of the data is consistent with the observed statistics, for example, whether the data can be generated from a Gaussian distribution, or whether a certain statistical model of a random graph will produce a graph with observed characteristics such as the power law of node degrees, etc.

A variety of approaches are used in practice to address many of these questions. They include the probabilistic modeling approach (with a well-defined statistical model), the non-probabilistic approach based on optimization, or simply a procedure that tries to find desired structures (that may or may not rely on optimization). For example, a mixture model can be used as a statistical model for addressing the clustering problem. With a mixture model, in order to generate each data point, one first generates its mixture component, then generates the observation according to the probability distribution of the mixture component. Hence the statistical approach requires a probabilistic model that generates the data—a so-called generative model. By comparison, the k-means algorithm assumes that the data are in k clusters, represented by their centroids. Each data point is then assigned to the cluster whose centroid is closest. This is iterated, and the algorithm converges to a local minimum of an appropriate distance-to-center criterion. This approach does not hinge on a statistical model, but instead on a sensible optimization criterion. There are also valid clustering procedures that are not based on optimization or statistical models. For example, in hierarchical agglomerative clustering, one starts with each single data point as a cluster, and then iteratively groups the two closest clusters to form a larger cluster; this process is repeated until all data are grouped into a single cluster. Hierarchical clustering does not depend on a statistical model of the data nor does it attempt to optimize a criterion. Nevertheless, it achieves the basic goal of cluster analysis—to find partitions of the data so that points inside each cluster are close to one another but not close to points in other clusters. In a loose sense, it also builds a useful model for the data that describes similarity relationship among observations. However, the model is not detailed enough to generate the data in a probabilistic sense.

Statistical models in the unsupervised setting that focus on the underlying data-generation mechanism can naturally be studied under the Bayesian framework. In that case, one is especially interested in finding unobserved hidden information from the data, such as factors or clusters that reveal some underlying structure in the data. Bayesian methods are natural in this context because they work with a joint distribution, both on observed and unobserved variables, so that elementary probability calculations can be used for statistical inference. Massive data may contain many variables that require complex probabilistic models, presenting both statistical and computational challenges. Statistically one often needs to understand how to design nonparametric Bayesian procedures that are more expressive than the more traditional parametric Bayesian models. Moreover, in order to simplify the specification of the full joint probability distribution, it is natural to consider simplified relationships among the data such as with graphical models that impose constraints in the form of conditional independencies among variables. Computational efficiency is also a major challenge in Bayesian analysis, especially for massive data. Methods for efficient large-scale Monte Carlo simulation or approximate inference algorithms (such as variational Bayesian methods) become important for the success of the Bayesian approach.

Supervised Learning

Predictive modeling is referred to as supervised learning in the machine-learning literature. One has a response or output variable Y, and the goal is to build a function $f(X)$ of the inputs X for predicting Y. The response "supervises" the learning procedure, in that it determines when the method is doing well or not. Basic prediction problems involving simple outputs include classification (Y is a discrete categorical variable) and regression (Y is a real-valued variable).

Statistical approaches to predictive modeling can be generally divided into two families of models: generative models and discriminative models. In a generative model, the joint probability of X and Y is modeled; that is, $P(X|Y)$. The predictive distribution $P(Y|X)$ is then obtained via Bayes's theorem. In a discriminative model, the conditional probability $P(Y|X)$ is directly modeled without assuming any specific probability model for X. An example of a generative model for classification is linear discriminant analysis. Here one assumes in each class the conditional distribution is Gaussian (with common covariance matrix used for all classes); hence, the joint distribution is the product of a Gaussian density with a class probability. Its discriminative model counterpart is linear logistic regression, which is also widely used in practice. Logistic regression proposes a model for $P(Y|X)$ and is not concerned with estimating the distribution of X. It turns out that they both result in the same parametric representation for $P(Y|X)$, but the two approaches lead to different estimates for the parameters.

In the traditional statistical literature, the standard parameter estimation method for developing either a generative or discriminative model is maximum likelihood estimation (MLE), which leads to an optimization problem. One can also employ optimization in predictive modeling in a broader sense by defining a meaningful criterion to optimize. For example, one may consider a geometric concept such as a margin and use it to define an optimization criterion for classification that measures how well classes are separated by the underlying classifier. This leads to purely optimization-based machine learning methods (such as the support vector machine method for recognizing patterns) that are not based on statistical models of how the data were generated (although one can also develop a model-based perspective for such methods).

Another issue in modern data analysis is the prevalence of high-dimensional data, where a large number of variables are observed that are difficult to handle using traditional methods such as MLE. In order to deal with the large dimensionality, modern statistical methods focus on regularization approaches that impose constraints on the model parameters so that they can still be reliably estimated even when the number of parameters is large. Examples of such methods include ridge regression and the Lasso method for least-squares fitting. In both cases one adds a penalty term that takes the form of a constraint on the norm of the coefficient vector (L_2 norm in the case of ridge regression, and L_1 norm in the case of Lasso). In the Bayesian statistical setting, constraints in the parameter space can be regarded naturally as priors, and the associated optimization methods correspond to maximum a posteriori estimation.

In many complex predictive modeling applications, nonlinear prediction methods can achieve better performance than linear methods. Therefore, an important research topic in massive data analysis is to investigate nonlinear prediction models that can perform efficiently in high dimensions. Classical examples of nonlinear methods include nearest neighbor classification and decision trees. Some recent developments include kernel methods, random forests, and boosting.

Some practical applications require one to predict output Y with a rather complex structure. For example, in machine translation, input X is observed as a sentence in a certain language, and a corresponding sentence (translation) Y needs to be generated in another language. These kinds of problems are referred to as structured prediction problems, an active research topic in machine learning. Many of these complex problems can benefit from massive data, even without increasing the complexity of the underlying models. Nevertheless, additional computational challenges arise. Efficient leverage of massive data is an important research topic currently in structured prediction, and this is likely to continue for the near future.

Another active research topic is online prediction, which can be regarded both as modeling for sequential prediction and as optimization over massive data. Online algorithms have a key advantage in handling massive data, in that they do not require all data to be stored in memory. Instead, each time they are invoked, they look at single observations (or small batches of observations). One popular approach is stochastic gradient descent. Because of this advantage, these algorithms have received increasing attention in the machine learning and optimization communities. Moreover, from a modeling perspective, sequential prediction is a natural setting for many real-world applications where data arrive sequentially over time.

Computational Simulation

Agent-based models and system dynamic models are core modeling techniques for assessing and reasoning about complex socio-technical systems where massive data are inherent. These models require the fusion of massive data, and the assessment of said data, to set initial conditions. In addition, these models produce massive data, potentially comparable in size and complexity to real-world data. Two examples, used in epidemiology and biological warfare, are BioWar (Carley et al., 2006) and Episims (Eubank et al., 2004). Both simulate entire cities, and thus are both users and producers of massive data. BioWar, for example, generates data on who interacts with whom and when, who has what disease, who is showing which symptom(s) and where, and what they are doing at a given time; it updates this picture across a city for all agents in 4-hour time blocks. For these models, core challenges are identifying reduced-form solutions that are consistent with the full model, storing and processing data generated, fusing massive amounts of data from diverse sources, and ensuring that results are due to actual behavior and not tail constraints on long chains of data.

Network Modeling

In data analysis, increasingly researchers are using relational or network models to assess complex systems. Models of social interaction, communication, and technology infrastructure (e.g., power grids) are increasingly represented and assessed as time-varying probabilistic networks. It is increasingly common for such models to have millions of nodes. A core challenge includes generation of massive but realistic network data that match real data not just in size and density, but also in the distribution of core metrics, linkage to other networks, and attributes of nodes. A second core challenge centers on statistically assessing confidence in network metrics given different types and categories of network errors. Row-column dependencies in networks violate the assumptions of simple parametric models and have driven the development of nonparametric approaches. However, such approaches are often computationally intensive, and research on scalability is needed. Another core challenge is then how to estimate confidence without requiring the generation of samples from a full joint distribution on the network.

MODEL TUNING AND EVALUATION

Models are, by their nature, imperfect. They may omit important features of the data in either the structural or noise components, make unwarranted assumptions such as linearity, or be otherwise mis-specified. On the other hand, models that are over-specified, in terms of being richer than the data can support, may fit the training data exceptionally well but generalize poorly to new data. Thus, an important aspect of predictive modeling is performance evaluation. With modern methods, this often occurs in two stages: model tuning, and then the evaluation of the chosen model. Tuning is discussed first, followed by model evaluation.

The models that are fit to particular data are often indexed by a tuning parameter. Some relevant examples are the following:

- The shrinkage parameter in a Lasso or elastic-net logistic regression;
- The number of terms (trees) in a boosted regression model;
- The "cost" parameter in a support vector machine classifier, or the scale parameter of the radial kernel used;
- The number of variables included in a forward stepwise regression; and
- The number of clusters in a prototype model (e.g., mixture model).

The process of deciding what model type to work with remains more art than science. In the massive data context, computational considerations frequently drive the choice. For example, in sequence modeling, a conditional random field may lead to more accurate predictions, but a simple hidden Markov model may be all that is feasible. Similarly, a multilevel Bayesian model may provide an elegant inferential framework, but computational constraints may lead an analyst to a simpler linear model. For many applications, a model-complexity ladder exists that provides the analyst with a range of choices. For example, in the high-dimensional classification context, the bottom rung contains simple linear classifiers and naive

Bayes. The next rung features traditional tools, such as logistic regression and discriminant analysis. The top rungs might feature boosting approaches and hierarchical nonparametric Bayesian methods. Similarly in pharmaco-epidemiology, simple (and widely used) methods include disproportionality analyses based on two-by-two tables. Case-control and case-crossover analyses provide a somewhat more complex alternative. High-dimensional propensity scoring methods and multivariate self-controlled case series are further up the ladder. Ultimately the appropriate rung on the ladder must depend on the signal-to-noise ratio. Presumably there is little point in fitting a highly complex model to data with a low signal-to-noise ratio, although little practical guidance currently exists to inform the analyst in this regard.

Ideally the family of models has been set up so that a tuning parameter orders the models in complexity. All the examples given above are of this kind. Complexity is generally understood here as the "effective dimension" of the model. The complexity is increased in an attempt to remove any systematic bias in the model. However, higher complexity also means that the model will fit the training data more closely, and there is a risk of over-fitting. The idea is to fit a sequence or path of models—one for each value of the tuning parameter—and evaluate the performance of each against a set of held-back "validation" data. Then one picks the position on the path with the best validation performance. Because the family is ordered according to complexity, this process determines the right complexity for the problem at hand.

A substantial body of knowledge now exists about complexity trade-offs in modeling. As mentioned previously, more complex models can over-fit data and provide poor predictions. These trade-offs, however, are poorly understood in the context of massive data, especially with non-stationary massive data streams. It is also important to note that statistical model complexity and computational complexity are distinct. Given a model with fixed statistical complexity and for a fixed out-of-sample accuracy target, additional data allow one to estimate a model with more computational efficiency. This is because in the worst case a computational algorithm can trivially subsample the data to reduce data size without hurting computation, but some algorithms can utilize the increased data size more efficiently than simple subsampling. This observation has been discussed in some recent machine-learning papers. However, in nonparametric settings, one may use models whose complexity grows with increasing data. It will be important to study how to grow model complexity (or shift models in the non-stationary setting) in a computationally efficient manner.

There are other good reasons for dividing the model-building process into the two stages of fitting a hierarchical path of models, followed by performance evaluation to find the best model in the path. One of these reasons is that typically the model fitting uses a "convenient" loss function that can be optimized with numerical algorithms—for example, a loss function that is convex and/or differentiable. This can be relaxed during the second stage, which can use whatever figure of merit has most meaning for the specific application, and it need not be smooth at all. Examples are misclassification cost or the "F1" measure, or some tailor-made cost function involving real prices and other factors.

In most cases this tuning is performing some kind of trade-off between bias and variance; essentially, deciding between under- and over-fitting of the training data. Thus, once the best model has been chosen, its predictive performance should be evaluated on a different held-back test data set, because the selection step can introduce bias. Ideally, then, the following three separate data sets will be identified for the overall task:

- *Training data.* These data are used to fit each of the models indexed by the tuning parameter. Typically one fits models by minimizing a smooth measure of training risk, perhaps penalized by a regularization term.
- *Validation data.* These data are used to evaluate the performance of each of the models fit in the previous step; that is, to evaluate the prediction performance. One then chooses a final model from this list (which can be a single model or a weighted combination of several).
- *Test data.* A final data set is often reserved to evaluate the chosen model, because the previous step can be viewed as "fitting" the validation data.

Model validation refers to using the validation data to evaluate a list of models. A plot of model prediction error versus tuning parameter values can be revealing. To see this, assume that as the tuning parameter increases, the model complexity increases. Then two general scenarios tend to occur in practice:

- *The prediction error initially decreases, bottoms out, and then starts to increase.* The decrease occurs because the early models are too restrictive or biased. Eventually the increase is because the models are fitting the noise in the training data, which adds unwanted variance that increases the prediction error when the model is applied to new data.
- *The error decreases and slowly levels out, never really increasing.* This often occurs in a data-rich scenario (many observations compared to variables). Here, the tuning allows one to fit a model that is sufficiently large to capture the structure in the data, while avoiding overly massive models that might strain resources.

For scenarios that are not data rich, or if there are many more variables than observations, one often resorts to K-fold cross-validation (Hastie et al., 2009). For this model-tuning task, the training data are randomly divided into K equal-sized chunks (K = 10 is a popular choice). One then trains on all but the kth chunk and evaluates the prediction error on the kth chunk. This is done K times, and the prediction-error curves are averaged.

One could also use cross-validation for the final task of evaluating the test error for the chosen model. This calls for two nested layers of cross-validation.

With limited-size data sets, cross-validation is a valuable tool. Does it lose relevance with massive data sources? Is the bias/variance trade-off even relevant anymore? Can one over-train? Or is one largely testing whether one has trained enough? If the data are massive because the number of variables is large in comparison to the number of observations, then validation remains essential (see the discussion of false discovery in the opening part of the section below on "Challenges"). If both the number of variables and the number of observations are large (and growing), then one can still overfit, and here too regularization and validation continue to be important steps. If there are more than enough samples, then one can afford to set aside a subset of the data for validation. As mentioned before, variance is typically not an issue here—one is determining the model complexity needed to accomplish the task at hand. This can also mean determining the size of the training set needed. If models have to repeatedly be fit as the structure of the data changes, it is important to know how many training data are needed. It is much easier to build models with smaller numbers of observations.

Care must be taken to avoid sample bias during the validation step, for the following reasons:

- The sampling procedure may itself be biased (non-random),
- Dynamics change in time-aware applications, and
- Data-gathering may involve adversarial actions.

An example of the last of these items is spam filtering, where the spammer constantly tries to figure out the algorithms that filter out spam. In situations like these, one may fit models to one population and end up testing them and making predictions on another.

With massive data it is often essential to sample, in order to produce a manageable data set on which algorithms can run. A common, simple example is where there is a rare positive class (such as "clicked on an ad") and a massive negative class. Here, it is reasonable to sample the positive and negative examples at different rates. Any logistic regression or similar model can be fit on the stratified sample and then post-facto corrected for the imbalance. This can be done on a larger scale, balancing for a variety of other factors. While it makes the modeling trickier, it allows one to work with more modest-sized data sets. Stratified sampling of this kind is likely to play an important role in massive data applications. One has to take great care to correct the imbalance after fitting, and account for it in the validation. See Chapter 8 for further discussion of sampling issues.

With massive data streams, it appears there may be room for novel approaches to the validation process. With online algorithms, one can validate before updating—i.e., evaluate the performance of the current model on a new data point, then update the model. If the complexity of the model is governed by the number of learning steps, this would make for an interesting adaptive learning algorithm. There is a large literature on this topic; indeed, online learning and early stopping were both important features in neural networks. (See, for example, Ripley, 1996, and Bishop, 1995.)

Once a model has been selected, one often wants to assess its statistical properties. Some of the issues of interest are standard errors of parameters and predictions, false discovery rates, predictive performance, and relevance of the chosen model, to name a few.

Standard error bars are often neglected in modern large-scale applications, but predictions are always more useful with standard errors. If the data are massive, these can be negligibly small, and can be ignored. But if they are not small, they raise a flag and usually imply that one has made a prediction in a data-poor region.

Standard error estimates usually accompany parameter and prediction estimates for traditional linear models. However, as statistical models have grown in complexity, estimation of secondary measures such as standard errors has not kept pace with prediction performance. It is also more difficult to get a handle on standard errors for complex models.

The bootstrap is a very general method for estimating standard errors, independent of the model complexity. It can be used to estimate standard errors of estimated parameters or predictions at future evaluation points. The committee's description focuses on the latter. With the bootstrap method, the (entire) modeling procedure is applied to many randomly sampled subsets of the data, and a prediction is produced from each such model. One then computes the standard deviation of the predictions derived from each sample. The original bootstrap takes samples of size N (the original size) with replacement from the training data, which may be infeasible in the setting of massive data. A recent proposal, known as the "bag of little bootstraps," has shown how to achieve the effect of the bootstrap while working with relatively small subsamples of the data (Kleiner et al., 2012).

The bootstrap is also the basis for model averaging in random forests, and in this context is somewhat similar to certain Bayesian averaging procedures.

Massive data streams open the door to some interesting new modeling paradigms that need to be researched to determine their potential effectiveness and usefulness. With parallel systems one could randomize the data stream and produce multiple models and, hence, predictions. These could be combined to form average predictions, prediction intervals, standard errors, and so on. This is a new area that has not been studied much in the literature.

CHALLENGES

Building effective models for the analysis of massive data requires different considerations than building models for the kinds of small data sets that have been more common in traditional statistics. Each of the topics outlined above face new challenges when the data become massive, although having access to much more data also opens the door to alternative approaches. For example, are missing data less of an issue, because we have so much data that we can afford to lose some measurements or observations? Can we simply discard observations with missing entries? While the answers are probably "no" to both these questions, it may well be that better strategies can be developed to deal with missing data when the source is less limited.

Likewise, does dirty (mislabeled or incorrect) data hurt as much if we have a large amount of it? Can the "dirt" get absorbed into the noise model and get washed out in any analysis? Again, the answer is probably "no" in general, but it may well be that procedures that were deemed statistically inefficient for small data might be reasonable with massive data. The committee notes that, in general, while sampling error decreases with increasing sample size, bias does not—big data does not help overcome bad bias.

Because massive amounts of observational data are exposed to many sources of contamination, sometimes through malicious intervention, can models be built that self-protect against these various sources? "Robust" regression models protect against outliers in the response, but this is just one source of contamination. Can the scope of these models be enlarged to cover more and different sources of contamination?

The literature currently frames model selection around a bias/variance trade-off. Is this still relevant for massive data—that is, is variance still an issue? In some cases, such as when there are many variables p per observation N ($p > N$; "wide" data), it will be. But in cases for which $N > p$ ("tall" data), the issue is less clear. If the models considered involve combinations of the p variables (interactions), then the numbers of such combinations grow rapidly, and many of them will be impacted by variance. In general, how will model selection change to reflect these issues? One suggestion is to find a least complex model that explains the data in a sufficient manner.

Is cross-validation relevant for massive data? Cross-validation has to live with correlations between estimates from different subsets of the data, because of overlap. This has an impact on, for example, standard error estimates for predictions (Markatou et al., 2005). Is one better off using independent subsets of the data to fit the model sequences, or some hybrid approach?

The bootstrap is a general tool for evaluating the statistical properties of a fitted model. Is the bootstrap relevant and feasible (e.g., will is scale) for massive data? The "bag of little bootstraps" exploits parallelism to allow the bootstrap to be applied on the scale of terabytes, but what about larger scales? Are there better and more efficient ways to achieve the effect of the bootstrap?

In many wide-data contexts, false discoveries—spurious effects and correlations—are abundant and a serious problem. Consider the following two cases:

- Gene-expression measurements are collected on 20,000 genes on 200 subjects classified according to some phenotype.
- Measurements of 1 million single-nucleotide polymorphisms are made on a few thousand case-control subjects.

Univariate tests at the $p = 0.01$ level would deliver 200 false discoveries in the first case, 10,000 in the second. As a result, traditional p values are less useful. Methods to control the false discovery rate (FDR) have been developed that take account of the abundant correlations between the many variables, but these are largely confined to univariate screening methods. There is a lot of room for development here in the context of massive data sets.

Finally, with massive data sets (large N [and possibly p]), it may be easy to find many "statistically significant" results and effects. Whether these findings have substantive relevance, however, becomes an important question, and statistically significant correlations and effects must be evaluated with subject-matter knowledge and experience.

Following are some highlighted challenges related to model building with massive data.

The Trade-Off Between Model Accuracy and Computational Efficiency

As was discussed in Chapter 6, computational complexity is an important consideration in massive data analysis. Although complex models may be more accurate, significantly more computational resources are needed to carry out the necessary calculations. Therefore determining the appropriate trade-off can be a challenging problem.

Even without more sophisticated models, if existing models can be adapted to handle massive data, then the mere availability of large amount of data can help to dramatically increase the performance. The challenge here is to devise computationally efficient algorithms for (even simple) models that can benefit from large data sets. An example is Google's translation and speech recognition systems, which while largely using models that have existed for many years, have significantly improved performance owing to the availability of large amounts of data gathered from the Internet and other sources. For example, translations of many phrases that cannot be found in small data sets can be easily found in new larger-scale and continuously updated repositories that contain a huge amount of text together with corresponding translations (e.g., news articles on the Web). It is thus not surprising that relatively simple translation models that map phrases in the source language to translated phrases in the target language can benefit significantly from more data. This example shows that a major challenge and opportunity in massive data analysis is to develop relatively simple models that can benefit from large amounts of data in a computationally efficient manner. Such models can be more effective than more complex models if the latter cannot be easily adapted to handle massive data sets.

Feature Discovery

Although simple regularized linear models such as the Lasso have been extensively studied in recent years, massive data analysis presents additional challenges both computationally and statistically. One may try to address the computational challenges in linear models with online algorithms. Nevertheless, as pointed out in the previous subsection, the availability of massive data means that there is a significant opportunity to build ever more complex models that can benefit from massive data, as long as this can be done efficiently.

One approach is to discover important nonlinear features from a massive data set and use linear models with the discovered features. This approach addresses the need for adding nonlinearity by separating the nonlinear feature discovery problem from the linear-model parameter estimation problem. The problem of efficient linear-model estimation has been extensively investigated in the literature, and it is still an important research topic. Therefore, developing the capability to discover nonlinear features is an important challenge that allows a data analyst to benefit from massive data under the simpler linear model framework.

The area of "deep learning" (Bengio, 2009) within machine learning focuses on features composed of multiple levels of nonlinear operations, such as in neural nets with many hidden layers or in complicated propositional formulae re-using many sub-formulae. Searching the parameter space of deep architectures is a difficult task, but learning algorithms such as those for deep belief networks have recently been proposed to tackle this problem with notable success, beating the state-of-the-art in certain areas.

A current important application area where feature discovery plays a central role concerns prediction of health-care events. For example, a recent data mining competition focuses on using insurance claims data to predict hospitalization[2]: Given extensive historical medical data on millions of patients (drugs, conditions, etc.), identify the subset of patients that will be hospitalized in the next year. Novel approaches are required to construct effective features from longitudinal data.

Ranking and Validation

A major goal in massive data analysis is to sieve through huge amounts of information and discover the most valuable information. This can be posed as statistical ranking problem, where the goal is to rank a set of items so that what is ranked on top contains the most important (or relevant) items. In general the more relevant the items that are placed on top, the higher the quality of the ranking algorithm. For example, FDR may be regarded as a ranking metric because it measures false negatives in top-ranked selections, and this is related to the concept of precision in the information retrieval literature.

One challenge is to design statistically sensible metrics to measure ranking quality and study statistical inference algorithms to optimize it (Duchi et al., 2012). The current search algorithms all use home-grown criteria of this kind.

Exploration in Evaluation

Some problems in massive data analysis require validation during exploration. An example is Internet advertising, where the goal is to serve ads and optimize revenue. In order to optimize the immediate reward, companies try to serve ads according to the current model. However, the models are under constant development, and parameters of the models are continually tested. Tang et al. (2010) report on some progress made in this area at Google. There is a big statistical challenge here in that, rather than evaluating one model at a time, one needs to be able to test an ever-changing set of models simultaneously, after which the effects of the parameters in the different models have to be teased apart. Just how to balance these operations is somewhat uncharted territory.

---

[2] Heritage Provider Network Health Prize website, available at http://www.heritagehealth prize.com.

Meta Modeling

With massive data there is an additional challenge of meta-modeling, the bringing together of multiple models each designed for different purposes into an operational whole. Currently the trend is to fuse the results from diverse models that utilize different parts of common data. Three difficulties that arise include the following:

- Ensuring that the common data are from the same spatial-temporal region,
- Building a meta-model of the way in which the diverse models operate, and
- Providing insight into the model that is relevant given the scale of the data being examined.

Meta modeling is related to the issue of data diversity. In many massive data applications, data contain heterogeneous data types such as audio, video, images, and text. Much of the model-building literature and the current model-building arsenal focus on homogeneous data types, such as numerical data or text. It remains a challenge to develop models that can satisfactorily handle multiple data types.

It is also known that combining many different models that focus on different aspects of the problem can be beneficial. One example is the 2009 Netflix competition that aimed to improve the prediction of individual movie preference (in order for Netflix to make appropriate movie recommendations to each user). A $1 million prize was awarded to the winning system, which was a combination of many different models. Another example is IBM's Jeopardy playing system, "Watson," which beat two human champions in a widely publicized television show. A key component of that system is an engine that combines many specialized algorithms (or models) where each can answer some specific types of questions more effectively than others. It is thus critical to develop an effective "executive" algorithm that can evaluate which of the proposed answers is more reliable and make choices, so that the overall system can provide a more accurate final answer than any individual component.

Tail Behavior Analysis

In traditional statistics, an event that has only, say, 0.1 percent probability of occurring may be safely regarded as a "rare event," and it can be ignored. (Although such events are considered in the robust statistics literature, they are studied there for a different reason with more specialized assumptions.) However, in the realm of massive data, these so-called "rare events" can occur sufficiently frequently that they deserve special attention. This means that in general the analysis of tail behavior becomes a key challenge in models of massive data.

One example of tail behavior analysis is outlier (or anomaly) detection. Another related issue is false discovery, discussed earlier.

Best Use of Model Output

As noted earlier, certain models (e.g., agent-based simulations) also generate massive data. Little theory exists to guide analyses of their output, and no standards exist for the data-to-model-to-data-to-metadata workflow. Basic research on infrastructure support for the generation, archiving, and searching of data-to-model-to-data is thus needed.

Large-Scale Optimization

It is clear that optimization plays an important role in model building. This is because traditional statistical models can lead to estimators (such as MLE) that solve an optimization problem; moreover, appropriately formulated optimization formulations (such as k-means) can in some ways be regarded as models themselves. Complex models involve more complex optimization formulations. It is thus important to investigate optimization formulations and methods that can handle massive data.

A particularly important direction is online optimization algorithms. In many massive data applications the underlying data-generation mechanism evolves over time. This is especially true with massive data streams. Although there are some existing dynamic modeling tools to handle non-stationary data, they concern themselves mostly with modest-sized data streams. Online model-building algorithms are often vital in such situations. Online methods try to address the time and space complexity simultaneously. The latter is achieved by not requiring all data to be in memory simultaneously, which alleviates data-storage requirement. However, it still requires the current state (represented by a set of parameters) of the underlying statistical model to be stored and updated. In particular, the online concept does not directly address the issue of how to efficiently store large statistical models. Therefore, it will be helpful to study space requirements in streaming applications by studying models that can be efficiently maintained/updated in the online setting with space constraints. A number of sketching and streaming algorithms deal with efficient storage and update of relatively simple statistics. What can be done for more complex statistical models? For example, when not all features can be stored in memory, linear classifiers remain problematic today. Although some streaming ideas seem applicable, the quality (both in practice and in theory) requires further investigation. Distributed optimization may be necessary for ultra-large-scale applications (Boyd et al., 2011).

Can online algorithms be adapted for model fitting to do model selection as well? That is, as new data are obtained and the fit is updated, can the tuning parameters also be guided and changed? How should this be done? See Auer et al. (2002) for some (theoretical) developments on this problem, and, for a broader overview, see Cesa-Bianchi and Lugosi (2006).

Models and System Architecture

In truly massive data analysis, a single machine will generally not be able to process all of the data. It is therefore necessary to consider efficient computational algorithms that can effectively utilize multiple processors. This computational consideration can play a significant role in deciding what models to use. For example, in a distributed computing environment with many computers that are loosely connected, the communication cost between different machines is high. In this scenario, some Bayesian models with parameters estimated using Monte Carlo simulation can relatively easily take advantage of the multiple machines by performing independent simulations on different machines. However, efficiently adapting an online learning algorithm to take advantage of distributed computational environment is more difficult. A key challenge is to investigate models and the associated computational methods that can easily take advantage of the computational power in multi-processor computing environments. Graphics processing units (GPUs) also show considerable promise for certain kinds of computations such as large-scale optimization.

Causal Modeling

Harnessing massive data to support causal inference represents a central scientific challenge. Key application areas include climate change, health-care comparative effectiveness and safety, education, and behavioral economics. Massive data open up exciting new possibilities but present daunting challenges. For example, given electronic health-care records for 100 million people, can we ascertain which drugs cause which side effects? The literature on causal modeling has expanded greatly in recent years. but causal modeling in massive data has attracted little attention.

REFERENCES

Auer, P., N. Cesa-Bianchi, and C. Gentile. 2002. Adaptive and self-confident on-line learning algorithms. *Journal of Computer and System Sciences* 64(1):48-75.

Bengio, Y. 2009. Learning deep architectures for AI. *Foundations and Trends in Machine Learning* 2(1):1-127. Available at http://dx.doi.org/10.1561/2200000006.

Bishop, C.M. 1995. *Neural Networks for Pattern Recognition.* Clarendon Press, Oxford, U.K.

Boyd, S., N. Parikh, E. Chu, B. Peleato, and J. Eckstein. 2011. Distributed optimization and statistical learning via the alternating direction method of multipliers. *Foundations and Trends in Machine Learning* 3(1):1-122.

Carley, K.M., D. Fridsma, E. Casman, A. Yahja, N. Altman, L.-C. Chen, B. Kaminsky, and D. Nave. 2006. BioWar: Scalable agent-based model of bioattacks. *IEEE Transactions on Systems, Man and Cybernetics-Part A* 36.2:252-265.

Cesa-Bianchi, N., and G. Lugosi. 2006. *Prediction, Learning and Games.* Cambridge University Press, New York, N.Y.

Dasu, T., and T. Johnson. 2003. *Exploratory Data Mining and Data Cleaning.* Volume 479. Wiley-Interscience, Hoboken, N.J.

Duchi, J., L. Mackey, and M.I. Jordan. 2012, Submitted. The asymptotics of ranking algorithms. E-print available at http://arxiv.org/abs/1204.1688.

Eubank, S., H. Guclu, V.S. Anil Kumar, M. Marathe, A. Srinivasan, Z. Toroczkai and N. Wang. 2004. Modeling disease outbreaks in realistic urban social networks. *Nature* 429:180-184.

Hastie, T., R. Tibshirani, and J. Friedman. 2009. *Elements of Statistical Learning.* Springer, New York, N.Y.

Kleiner, A., A. Talwalkar, P. Sarkar, and M.I. Jordan. 2012, Submitted. A scalable bootstrap for massive data. E-print available at http://arxiv.org/abs/1112.5016.

Markatou, M., H. Tian, S. Biswas, and G. Hripcsak. 2005. Analysis of variance of cross-validation estimators of the generalization error. *Journal of Machine Learning Research* 6:1127-1168.

McCullagh, P. 2002. What is a statistical model? (with discussion). *Annals of Statistics* 30:1225-1310.

Ripley, B. 1996. *Pattern Recognition and Neural Networks.* Cambridge University Press, Cambridge, U.K.

Tang, D., A. Agarwal, and M. Meyer. 2010. Overlapping experiment infrastructure: more, better, faster experimentation. Pp. 17-26 in *Proceedings of the 16th ACM SIGKDD International Conference on Knowledge Discovery and Data Mining.* Washington, D.C.

Wainwright, M., and M.I. Jordan. 2003. Graphical models, exponential families and variational inference. *Foundations and Trends in Machine Learning* 1:1-305.

8

Sampling and Massive Data

Sampling is the process of collecting some data when collecting it all or analyzing it all is unreasonable. Before addressing why sampling still matters when massive amounts of data are available and what the new challenges are when the amount of data is massive, an overview of statistical sampling is provided below.

COMMON TECHNIQUES OF STATISTICAL SAMPLING

Random Sampling

In simple random sampling, everything of interest that could be sampled is equally likely to be included in the sample. If people are to be sampled, then everyone in the population of interest has the same chance to be in the sample. For that reason, a simple random sample gives an unbiased representation of the population.

Simple random sampling is usually straightforward to implement, but other kinds of sampling can give better estimates. Stratified random sampling partitions the population into groups called strata and then randomly samples within each group. If strata are less diverse than the population as a whole, then combining group-level estimates, such as group means, is better than estimating from the population without partitioning. The improvement due to partitioning is larger if the more heterogeneous groups are more heavily sampled. Often the groups and group sizes are chosen to optimize some criterion—such as constraining the mean squared error of an estimate—based on past data, theory about the processes generating the data, or intuition. Of course, a sampling strategy that is optimal for estimating one kind of parameter may be inefficient or even unusable for another, so there is often an informal compromise between optimal stratification and pure random sampling so that useful information can be obtained about a wider set of parameters. This kind of compromise is especially important when data are expensive to collect and may be re-used for unforeseen purposes.

There are many variants of random sampling beyond stratification. For example, random sampling may be applied in stages. For example, cluster or hierarchical sampling first randomly samples city blocks, apartment buildings, households, or other "clusters," and then randomly samples individuals within each sampled cluster. Panels that monitor changes in attitudes or health track a random sample of respondents over time, removing respondents from the panel when they have served for a fixed length of time, and replacing them with a new random sample of respondents. Only a fraction of the respondents enter or leave the panel at the same time, so the panel is always a mix of different "waves" of respondents. Nielsen runs a panel of households for television viewing, Comscore has a panel for Web browsing, and the U.S. government has many panels, such as the National Longitudinal Studies run by the Bureau of Labor Statistics. Other examples of staged sampling can be found in environmental science and ecology. For example, points or small regions are randomly sampled, and then data are taken along random directions from a landmark, such as the center of the sampled region. Of course, random sampling methods can be combined, so panels can be selected with stratification, for example.

Sampling may also be adaptive or sequential, so that the sampling rule changes according to a function of the observations taken so far. The goal is to over-sample regions with "interesting" data or more-variable data to ensure that the estimates in those regions are reliable. For example, the number of observations taken in a region may not be fixed in advance but instead depend on the data already observed in the region or its neighbors. In sequential sampling, the population to be sampled (called the *sampling frame*) may change over time, as happens when sampling a data stream.

Randomly sampling a data stream may mean producing a set of observations such that, at any time, all observations that have occurred so far are equally likely to appear in the sample. Vitter (1985) called this reservoir sampling, although his algorithm was earlier known as the Fisher-Yates shuffling algorithm (Fisher and Yates, 1938). The basic idea behind reservoir sampling is that the probability that the next item seen is sampled depends on the number of items seen so far, but not otherwise on their values.

Random sampling of data streams has two major disadvantages: the number of unique values grows, which requires more and more storage, and the newly observed values, which are often the most interesting, are given no more weight than the old values. Gibbons and Mattias (1998) consider fixed-buffer-length sampling schemes that replace an element in a buffer (the current fixed-length sample) with a value that is not in the sample with a probability that depends on the buffer counts so far. The basic idea is that a newly observed value that recurs, and so is "hot," will eventually be included in the sample. Variants of this scheme use exponentially weighted moving averages of the buffer probabilities and include a buffer element for "other," which tracks the probability that the buffer does not include a recent item.

In choice-based and case-based sampling, whether for fixed populations or data streams, the probability of selection depends on an outcome like disease status. Case-based sampling is necessary when an outcome is so rare that simple random sampling would likely produce too few positive cases and an overwhelming number of negative cases. It is used in applications as diverse as medicine, ecology, and agriculture. For example, Suter et al. (2007) matched grassland plots with the poisonous grassland weed *Senecio jacobaea* and neighboring plots without *S. jacobaea* to determine environmental conditions conducive to the growth of *S. jacobaea*.

Event-based sampling, in which data are collected only when a signal exceeds a threshold or when an alarm sounds, is yet another kind of adaptive sampling. It is commonly used for analyses from engineering and Earth and planetary sciences for which storing all the data would be too costly or impractical. The threshold used depends on scientific knowledge about the size of interesting events. For example, the data rate on detectors in high-energy physics is so extreme that thresholds in hardware only allow 1 in 10 million events to pass to the next stage of the data system. The surviving data are then hierarchically sampled and filtered for further processing.

Size-biased sampling is similar: the probability of selection depends, either intentionally or not, on a measure of size, such as the duration of hospital stay or revenues of a corporation, that is related to the outcome. Size bias that is intentional can be removed during estimation and model fitting by weighting the data; unintentional size bias requires more specialized statistical analyses to remove (e.g., Vardi, 1982).

Specialized sampling techniques have evolved in ecology and evolutionary and environmental biology, and some of these are applied in large-scale applications. For example, estimation of wildlife population sizes has long used capture-recapture. The key idea is to capture animals, tag them, and then recapture the same population. The fraction of previously unseen animals in the recapture provides information about the population size. Capture-recapture designs can be quite elaborate and are now widely used in epidemiology.

There is also a growing literature on random sampling for simple networks and graphs; for example, Handcock and Gile (2010). The network is represented as a binary matrix where element (i, j) is 1 if that pair of nodes is sampled, and 0 if not. Random sampling then chooses one such binary matrix randomly. For example, a node i may be chosen by simple random sampling and then all the nodes j that interact with i are included in the sample. Such sampling preserves some relationships, but it destroys many others. Sampling on networks is still in its infancy, and there are obvious opportunities for stratification, clustering, and adaptation (such as link tracing), with different consequences for estimation.

There are many more principled ways to sample data randomly. In theory, random sampling is easy, but in practice it is fraught with unforeseen complications, often due to a poor match between the sampling design and the structure and noise of the data.

Sampling according to known probabilities, whether equal or not, or whether all-at-once or sequentially, is called random sampling. Stratified sampling, case-based sampling, adaptive sampling, and length-biased sampling are all examples of random sampling. They all share the premise that the probability model that defines the sampling plan can be "unwound" to estimate a parameter of interest for the population at large, even if some observations, by design, were more likely to be sampled than others. Unequal weights give intentional sampling bias, but the bias can be removed by using the sampling weights in estimation or by estimating with regression or other models, and the bias allows attention to be focused on parts of the population that may be more difficult to measure. Finally, whether simple or not, random sampling leads to valid estimates of the reliability of the estimate itself (its uncertainty).

It is important to note that the analysis goals drive the sampling. Consider for example a repository of Web usage data. Sampling of search queries—probably by country, language, day of week, and so on—may make sense if the goal is to measure the quality of search results. On the other hand, if the goal is to measure user happiness, then one should sample across users, and perhaps queries within users. Or, it might be appropriate to sample user sessions, which are defined by nearly uninterrupted periods of activity for a user. At the other extreme, the sampling unit might be low-level timing events within queries. The decision about how to sample should follow the decision about what to sample.

Non-Random Sampling

Unfortunately, random sampling is not always practical. Members of stigmatized populations, like intravenous drug users, may hide the fact that they are part of that population, so the sampling frame is unknown. Some populations, like that of people with particular expertise or perspectives, may be difficult to identify. Snowball sampling, proposed by Goodman

(1961), starts with a set of seeds that are known to belong to the hidden group. The seeds are asked to identify additional members of the group, then the newly identified members are asked to identify other group members, and so on. This process produces a sample without knowing the sampling frame. Goodman showed that if the initial seeds are random, then it is possible under some conditions to estimate relationships between people from snowball samples. Of course, there is a danger that the population is not well mixed, so only a non-representative sample can be reached from the initial seed. In contrast, random sampling is unbiased in the sense that two people in the same stratum, for example, have the same chance of being in the sample, even if their characteristics beyond those used for stratification are very different.

Heckathorn (1997, 2002) improved snowball sampling by adding more structure to sample recruitment. Each recruit at each stage is given a set number of coupons, a constraint that controls the rate of recruitment in sub-groups (as in stratification), reduces the risk of sampling only a dominant subgroup, and makes it easier to track who recruited whom and how large a social circle each recruit has. This respondent-driven sampling (RDS) has largely replaced snowball sampling. RDS is not random sampling because the initial seeds are not random, but some have suggested that the mixing that occurs when there are many rounds of recruitment or when the recruits added at later stages are restricted to those that have been given exactly one coupon induces a pseudo-random sample. However, difficulty in recruiting initial seeds may still lead to bias, and there is controversy about the reliability of estimates based on RDS even after correcting for the size of the networks of the initial seeds (e.g., Goel and Salganik, 2010). Finally, note that both RDS and snowball sampling often focus on estimation for an outcome in a population that is reached through a social graph rather than estimation of properties of the graph itself.

Sparse Signal Recovery

There are alternatives to sampling when collecting all the data is impractical, especially in applications in the physical sciences and engineering. In sparse signal recovery, a data vector x is multiplied by a matrix M, and only the resulting $y = Mx$ is retained. If x has only a few large components, then the matrix M can have considerably fewer rows than columns, so the vector y of observations is much smaller than the original data set without any loss of information. This kind of data reduction and signal recovery is known as compressed sensing, and it arises in many applications, including analog-to-digital conversion, medical imaging, and hyperspectral imaging, with a variety of different features that depend on the application (Candès and Tao, 2005; Donoho, 2006).

In signal and image processing, sparse recovery is done physically. For example, analog-to-digital converters sample analog signals, quantize the samples, and then produce a bit-stream. Hyperspectral cameras use optical devices to acquire samples that are then digitized and processed off-line. Streaming algorithms employ software-based measurement schemes.

Group testing is a special case of sparse signal recovery. Here the observations are taken on groups or pools of respondents, and only the combined response for the group is observed. Dorfman (1943), for example, combined blood samples from groups of army recruits to detect if any of them had syphilis. Because syphilis was uncommon, most groups would be negative, and no further testing was needed.

Sampling for Testing Rather Than Estimation

The discussion so far has focused on sampling for estimation, but it is also common to sample in order to test whether a set of factors and their interactions affect an outcome of interest. The factors under test are called *treatments*, and the possible values of a treatment are called *levels*. For example, two treatments might be "image used in ad campaign" and "webpage where the ad is shown." The levels of the first treatment might be the various images shown to the user, including the null value of "no ad," which is called the control. Often only a few levels of each treatment are tested. The decision about which test units get which combination of levels of the treatments is known as experiment design. Designed experiments go back at least as far as the 1700s, but Fisher's book *The Design of Experiments* (1935) was the first to lay out principles for statistical experimentation.

The basic premise of experiment design is that it is much more efficient and informative to apply several treatments to the same unit under test, thus testing the effect of several treatments simultaneously, than it is to apply only one experimental factor to each test unit. That is, instead of testing whether the different levels of treatment A give different results on average, and then testing whether the different levels of treatment B give different results on average, both A and B are tested simultaneously by assigning every test unit a level of A and a level of B.

There are many kinds of experiment designs, just as there are many sampling designs. Perhaps the most common are fractional factorials that allocate test units to combinations of levels of treatment factors when the number of treatment levels plus the number of treatment interactions of interest exceed the number of units available for testing. In that case, various criteria are optimized to choose which combinations of treatment levels to test and how many units to test at each combination. Classical combinatorial theory is often used to find the assignment of treatments to experimental units that minimize the expected squared error loss (e.g., Sloane and Hardin, 1993).

Test units are often a simple random sample from the population of interest, and they are often randomly assigned levels of the factors to be tested. However, test units can be blocked (stratified) to control for confounding factors that are not under test, such as age. For example, test units might be stratified by country when testing the effects of different ad images and websites. Then comparisons are made within blocks and combined over blocks. Sequential testing and adaptive testing (e.g., multi-armed bandits) are also common in some areas, such as in medicine. These designs change the fraction of test units assigned to each combination of treatment factors over time, eliminating treatment levels as soon as they are unlikely to ever show a statistically significant effect. These ideas also extend to online experiments, where tests of different treatments are started and stopped at different times. That is, treatments A, B, and C may be tested on day 1 and treatments A, B, and D tested on day 2. If standard design principles are followed, then the effects of the different treatments and their interactions can be teased apart (Tang et al., 2010).

Random sampling can be used not only to assign units to factors but also to evaluate which, if any, of the treatment factors and their interactions are statistically significant. As a simple example, suppose there is only one factor with two levels A and B, m units are assigned to A and n to B, and the goal is to decide if the mean outcome under A is different from the mean outcome under B after observing the difference D_{obs} of sample means. If the factor has no effect, so that there is no difference in A and B, then the A and B labels are meaningless. That is, D_{obs} should look like the difference computed for any random re-labeling of the units as m As and n Bs. The statistical test thus compares D_{obs} to the difference of sample means for random divisions of the observed data into A and B groups. If D_{obs} is larger or smaller than all but a small fraction of the mean differences obtained by random re-labeling, then there is statistical evidence that A and B have different effects on the mean outcome. On the other hand, if D_{obs} is not in the tail of the re-labeled differences, then one cannot be sure that the difference between A and B is meaningful. Such tests are called randomization or permutation tests. Tests that re-label a set of $m+n$ test units drawn with replacement from the original data are called bootstrapped tests; these behave similarly. Randomization and bootstrapped tests are nearly optimal in finite samples under weak conditions, even if the space of possible re-labelings is only sampled. In other words, valid tests can be constructed by approximating the sampling distribution of the test statistic with random re-sampling of the data.

Finally, it is not always possible to assign units to treatments randomly. Assigning students to a group that uses drugs to measure the effect of drug use on grades would be unethical. Or people may be assigned randomly to a treatment group but then fail to use the treatment. For example, a non-random subset of advertisers may be offered a new tool that should make it easier to optimize online ad campaigns, and a non-random subset of those may choose to use it. Or the treatment group may be identified after the fact by mining a database. Even though the treatment group is not random, valid tests can often be based on a random set of controls or comparison group, perhaps stratifying or matching so that controls and "treated" were similar before treatment. Testing without randomization or with imperfect randomization falls under the rubric of observational studies.

CHALLENGES WHEN SAMPLING FROM MASSIVE DATA

Impressive amounts of data can now be collected and processed. But some tasks, like data visualization, still require sampling to tame the scale of the data, and some applications require sampling and testing methods that are beyond the state of the art. Following are a few examples of the latter.

Data from Participatory Sensing, or "Citizen Science"

There are billions of cell phones in the world. In participatory sensing, volunteers with cell phones collect location-tagged data either actively (taking images of overflowing trash cans, for example) or passively (reporting levels of background noise, pollutants, or health measurements like pulse rates). In principle, simple random sampling can be used to recruit participants and to choose which active devices to collect data from at any time and location, but, in practice, sampling has to be more complex and is not well-understood. Crowdsourcing, such as the use of micro-task markets like Amazon's Mechanical Turk, raise similar sampling questions (see Chapter 9). Challenges, which result from the non-stationary nature of the problem, include the following:

- *Sample recruitment is ongoing.* The goal is to recruit (i.e., sample mobile devices or raters for micro-task markets) in areas with fast-changing signals (hot spots) or high noise or that are currently underrepresented to improve the quality of sample estimates. Those areas and tasks have to be identified on the fly using information from the current participants. Recruiting based on information from other participants introduces dependence, which makes both sampling implementation and estimation more difficult. There are analogies with respondent-driven sampling, but the time scales are much compressed.

- *Some incoming samples are best omitted.* This is, in a sense, the inverse of the sample recruitment challenge just mentioned: how to decide when to exclude a participant or device from the study, perhaps because it produces only redundant, noisy, spotty, or perverse data.
- *Participants will enter and leave the sensing program at different times.* This hints of panel sampling with informative drop-out, but with much less structure. Note that the problem is non-stationary when the goals change over time or the user groups that are interested in a fixed goal change over time.
- *Data need not be taken from all available devices all the time.* Randomly sampling available devices is one option, but that could be unwieldy if it requires augmenting information from one device with information from others. Here, the problems are intensified by the heterogeneous nature of the data sources. Some devices may be much more reliable than others. Moreover, some devices may give data in much different formats than others.
- *Sampling for statistical efficiency and scheduling for device and network efficiency have an interplay.* How should sampling proceed to minimize communication (e.g., battery) costs? There may be analogies with adaptive testing or experiment design.
- *Data obtained from participatory sensing and crowdsourcing is likely to be biased.* The mix of participants probably does not at all resemble a simple random sample. Those who are easiest to recruit may have strong opinions about what the data should show, and so provide biased information. Samples may need to be routinely re-weighted, again on the fly, with the weights depending on the purpose of the analysis. New ways to describe departures from random sampling may need to be developed. Past work on observational studies may be relevant here.

For recent references on sampling and sampling bias in crowdsourcing or citizen science, see, for example, Dekel and Shamir (2009), Raykar et al. (2010), and Wauthier and Jordan (2011).

Data from Social Networks and Graphs

Statistically principled sampling on massive graphs, whether static or dynamic, is still in its infancy. Sampling—no matter how fine its resolution—on graphs can never preserve all graph structure because sampling never preserves everything. But having some insight into some properties of a graph and its communities, and being able to characterize how reliable that insight is, may be better than having none at all. If so, then a set of design principles for sampling from graphs for different kinds of analyses is needed.

Variants of respondent-driven sampling have been applied to graphs by randomly sampling a seed set of nodes and then randomly sampling some of the nodes they are connected to, with the selection depending on a window of time for dynamic graphs. Sometimes new random seeds are added throughout the sampling. Other methods are based on random walks on graphs, starting from random nodes, perhaps adding new seeds at random stages. With either approach, there are many open questions about tailoring sampling methods to the analysis task. Just as with smaller problems, there is likely to be no one right way to sample, but rather a set of principles that guide sampling design. However, these principles are not yet in place.

Entirely new ways to sample networks may be needed to obtain dense-enough subgraphs to give robust estimates of graph relationships. Traditional network statistics are highly sensitive to missing data or broken relations (e.g., Borgatti et al., 2006). For example, a rank order of nodes can be radically different if as little as 10 percent of the links are missing. New ways to sample that preserve node rankings are needed, as is theoretical understanding of the biases inherent in a strategy for network sampling. In particular, sampling designs that account for the network topology are needed. There has been some initial work (e.g., Guatam et al., 2008), but this topic is still in its infancy.

Finally, the difficulties in sampling networks are compounded when the data are obtained by crowdsourcing or massive online gaming. They are further intensified when individuals belong to multiple social networks, maintained at different sites. Answering a question as seemingly simple as how many contacts does an individual have is fraught with technical difficulty.

Experiment design for social networks is even less explored. Here the unit under test is not a node but a connected set of nodes. Choosing equivalent sets of connected nodes that can be randomly assigned to the different levels of the treatments has only recently received attention (e.g., Backstrom and Kleinberg, 2011.) There is also a need for methods that allow principled observational studies on graphs. Aral et al. (2009) provide an early example that uses propensity scoring to find appropriate random controls when the treatment cannot be properly randomized.

Data from the Physical Sciences

The emergence of very large mosaic cameras in astronomy has created large surveys of sky images that contain 500 million objects today and soon will have billions. Most astronomers today use a target list from a large, existing imaging survey to select their favorite objects for follow-up observations. Because these follow-ups are quite expensive in terms of telescope time, various (mostly ad hoc) sampling strategies are applied when, for example, selecting galaxies for spectroscopic observations. Although it is relatively straightforward to create a sampling scheme for a single criterion, it is rare that a sample created for one purpose will not be later reused in another context. The interplay between often conflicting sampling criteria is poorly understood, and it is not based on a solid statistical foundation.

As a concrete example, both Pan-STARRS and the Large Synoptic Survey Telescope will provide deep multicolor photometry in co-added (averaged) images for several billion galaxies over a large fraction of the sky. This data set will provide an excellent basis for several analyses of large-scale structure and of dark energy. However, unlike the Sloan Digital Sky Survey (SDSS), these surveys will not have a spectroscopic counterpart of comparable depth. The best way to study structure, then, is to split the galaxy sample into radial shells using photometric redshifts—i.e., using the multicolor images as a low-resolution spectrograph. The most accurate such techniques today require a substantial "training set" with spectroscopic redshifts and a well-defined sample selection. No such sample exists today, nor is it clear how to create one. It is clear that its creation will require major resources (approximately 50,000-100,000 deep redshifts) on 8- to 10-m class telescopes.

Given the cost of such an endeavor, extreme efficiency is needed. Special care must be taken in the selection of this training set, because galaxies occupy a large volume of color space, with large density contrasts. Either one restricts galaxy selection to a small, special part of the galaxy sample (such as luminous red galaxies from the SDSS), or one will be faced with substantial systematic biases. A random subsample will lack objects in the rare parts of color space, while an even selection of color space will under-sample the most typical parts of color space. The optimum is obviously in between. A carefully designed stratified sample, combined with state-of-the-art photometric redshift estimation, will enable many high-precision cosmological projects of fundamental importance. This has not been attempted at such scale, and a fundamentally new approach is required.

Time-domain surveys are just starting, with projected detection cardinalities in the trillions. Determining an optimal spatial and temporal sampling of the sky (the "cadence") is very important when one operates a $500 million facility. Yet, the principles behind today's sampling strategies are based on ad hoc, heuristic criteria, and developing more optimal algorithms, based on state-of-the-art statistical techniques, could have a huge potential impact.

As scientists query large databases, many of the questions they ask are about computing a statistical aggregate and its uncertainty. Running an SQL statement or a MapReduce task provides the "perfect" answer, the one that is based on including all the data. However, as data set sizes increase, even linear data scans (the best case, because sometimes algorithms have a higher complexity) become prohibitively expensive. As each data point has its own errors, and statistical errors are often small compared to the known and unknown systematic uncertainties, using the whole data set to decrease statistical errors makes no sense. Applying an appropriate sampling scheme (even incrementally) to estimate the quantities required would substantially speed up the response, without a loss of statistical accuracy. Of course, scientific data sets often have skewed distributions; not everything is Gaussian or Poisson. In those cases, one needs to be careful how the sampling is performed.

REFERENCES

Aral, S., L. Muchnik, A. Sundararajan. 2009. Distinguishing influence based contagion from homophily driven diffusion in dynamic networks. *Proceedings of the National Academy of Sciences U.S.A.* 106:21544-1549.

Backstrom, L., and J. Kleinberg. 2011. Network bucket testing. Pp. 615-624 in *Proceedings of the 20th International World Wide Web Conference.* Association for Computing Machinery, New York, N.Y.

Borgatti, S., K.M. Carley, and D. Krackhardt. 2006. Robustness of centrality measures under conditions of imperfect data. *Social Networks* 28:124-136.

Candès, E., and T. Tao. 2005. Decoding by linear programming. *IEEE Transactions on Information Theory* 51:4203-4215.

Dekel, O., and O. Shamir. 2009. Vox Populi: Collecting high-quality labels from a crowd. Pp. 377-386 in *Proceedings of the 22nd Annual Conference on Learning Theory (COLT).* Available at http://www.cs.mcgill.ca/~colt2009/proceedings.html.

Donoho, D.L. 2006. Compressed sensing. *IEEE Transactions on Information Theory* 52: 1289-1306.

Dorfman, R. 1943. The detection of defective members of large populations. *The Annals of Mathematical Statistics* 14:436-440.

Fisher, R.A. 1935. *The Design of Experiments.* Macmillan, New York, N.Y.

Fisher, R.A., and F. Yates. 1938. *Statistical Tables for Biological, Agricultural and Medical Research,* 3rd ed. Oliver and Boyd, London. Pp. 26-27.

Gautam, D., N. Koudas, M. Papagelis, and S. Puttaswamy. 2008. Efficient sampling of information in social networks. Pp. 67-74 in *Proceeding of the 2008 ACM Workshop on Search in Social Media (SSM '08).* Association for Computing Machinery, New York, N.Y.

Gibbons, P.B., and Y. Matias. 1998. New sampling-based summary statistics for improving approximate query answers. Pp. 331-342 in *Proceedings of the 1998 ACM International Conference on Management of Data (SIGMOD).* Association for Computing Machinery, New York, N.Y.

Goel, S., and M.J. Salgonik. 2010. Assessing respondent-driven sampling. *Proceedings of the National Academy of Sciences U.S.A.* 107:6743-6747.

Goodman, L.A. 1961. Snowball sampling. *The Annals of Mathematical Statistics* 32:148-170.

Handcock, M.S., and Gile, K. 2010. Modeling social networks from sampled data. *The Annals of Applied Statistics* 4:5-25.

Heckathorn, D. 1997. Respondent-driven sampling: A new approach to the study of hidden populations. *Social Problems* 44:4174-199.

Heckathorn, D. 2002. Deriving valid population estimates from chain-referral samples of hidden populations. *Social Problems* 49:11-34.

Raykar, V.C., S. Yu, L.H. Zhao, G.H. Valadez, C. Florin, L. Bogoni, and L. Moy. 2010. Learning from crowds. *Journal of Machine Learning Research* 11:1297-1322.

Sloane, N.J.A., and R.H. Hardin. 1993. *Journal of Statistical Planning and Inference* 37: 339-369.

Suter, M., S. Siegrist-Maag, J. Connolly, and A. Lüscher. 2007. Can the occurrence of *Senecio jacobaea* be influenced by management practice? *Weed Research* 47:262-269.

Tang, D., A. Agarwal, D. O'Brien, and M. Meyer. 2010. Overlapping experiment infrastructure: More, better, faster experimentation. Pp. 17-26 in *Proceedings of the 16th Conference on Knowledge Discovery and Data Mining*. Association for Computing Machinery, New York, N.Y.

Vardi, Y. 1982. Nonparametric estimation in the presence of length bias. *The Annals of Statistics* 10:616-620.

Vitter, J.S. 1985. Random sampling with a reservoir. *ACM Transactions on Mathematical Software* 11:37-57.

Wauthier, F.L., and M.I. Jordan. 2011. Bayesian bias mitigation for crowdsourcing. Pp. 1800-1808 in *Proceedings of the Conference on Neural Information Processing System, Number 24*. Available at http://machinelearning.wustl.edu/mlpapers/papers/NIPS2011_1021.

9

Human Interaction with Data

INTRODUCTION

Until recently, data analysis was the purview of a small number of experts in a limited number of fields. In recent years, however, more and more organizations across an expanding range of domains are recognizing the importance of data analysis in meeting their objectives. Making analysis more useful to a wider range of people for a more diverse range of purposes is one of the key challenges to be addressed in the development of data-driven systems.

In many, perhaps most, scenarios today and in the near-term future, people are the ultimate consumers of the insights from data analysis. That is, the analysis of data is used to drive and/or improve human decision-making and knowledge. As such, methods for visualization and exploration of complex and vast data constitute a crucial component of an analytics infrastructure. The field of human-computer interactions has made great progress in the display and manipulation of complex information, but the increasing scale, breadth, and diversity of information provide continued challenges in the area.

People are not, however, merely consumers of data and data analysis. In many analytics usage scenarios, people are also the source (and often the subject) of the data being analyzed. Continuing improvements in network connectivity and the ubiquity of sophisticated communications and computing devices has made data collection easier, particularly as more activities are done online. Moreover, increasing network connectivity has been leveraged by a number of platforms that can allow people to participate directly in the data analysis process. Crowdsourcing is the term used to describe the harnessing of the efforts of individual people and groups to accomplish a larger task. Crowdsourcing systems have taken on many forms, driven largely by advances in network connectivity, the development of service-oriented platforms and application programming interfaces (APIs) for accomplishing distributed work, and the emergence of user-generated content sites (sometimes referred to as "Web 2.0") that include socially oriented and other mechanisms for filtering, vetting, and organizing content.

This chapter discusses several aspects of crowdsourcing that could contribute to extracting information from massive data. Crowdsourced data acquisition is the process of obtaining data from groups either explicitly— for example, by people deliberately contributing content to a website—or implicitly, as a side effect of computer-based or other networked activity. This has already been shown to be a powerful mechanism for tasks as varied as monitoring road traffic, identifying and locating distributed phenomena, and discovering emerging trends and events.

For the purposes of this report, a perhaps more interesting development in crowdsourcing is the involvement of people to aid directly in the analysis process. It is well known that computers and people excel at very different types of tasks. While algorithm developers continue to make progress in enabling computers to address tasks of greater complexity, there remain many types of analysis that can be more effectively done by people, even when compared to the most sophisticated computers and algorithms. Such analyses include deep language understanding and certain kinds of pattern recognition and outlier detection. Thus, there has been significant recent work and recognition of further opportunities in hybrid computer/human data analysis.

These trends are leading to the increased understanding of the role of people in all phases of the data processing lifecycle—from data collection through analysis to result consumption, and ultimately to decision making. The human dimension carries with it a new set of concerns, design constraints, and opportunities that must be addressed in the development of systems for massive data analysis. In addition, massive data calls for new approaches to data visualization, which is often used in exploratory data analysis. This chapter thus focuses on the various aspects of human interaction with data, with an emphasis on three areas: data visualization and exploration, crowdsourced data acquisition, and hybrid computer/human data analysis.

STATE OF THE ART

Data Visualization and Exploration[1]

Information visualization technologies and visual analytics processes have matured rapidly in the past two decades and continued to gain commercial adoption, while the research enterprise has expanded. Successful commercial tools include some that stand alone, such as Spotfire, Tableau, Palantir, Centrifuge, i2, and Hive Group, as well as some that are embedded in other systems, such as IBM ILOG, SAS JMP, Microsoft Proclarity, Google Gapminder, and SAP Xcelsius. In addition, open-source toolkits such as R, Prefuse, ProtoVis, Piccolo, NodeXL, and Xmdv support programmers. Academic conferences and journals in this area are active, and an increasing number of graduate courses are available.

At the same time, news media and web-based blogs have focused intense public attention on interactive infographics that deal with key public events such as elections, financial developments, social media impacts, and health care/wellness. Information visualizations provide for rapid user interaction with data through rich control panels with selectors to filter data, with results displayed in multiple coordinated windows. Larger displays (10 megapixels) or display arrays (100-1,000 megapixels) enable users to operate dozens of windows at a time from a single control panel. Large displays present users with millions of markers simultaneously, allowing them to manipulate these views by dynamic query sliders within 100 ms. Rapid exploration of data sets with 10 million or more records supports hypothesis formation and testing, enabling users to gain insights about important relationships or significant outliers.

An important distinction is often made between the more established field of scientific visualization and the emerging field of information visualization. Scientific visualizations typically deal with two-dimensional and three-dimensional data about physical systems in which questions deal with position—for example, the location of highest turbulence in the airflow over aircraft wings, the location and path of intense hurricanes, or the site of blockages in arteries. In contrast, information visualizations typically deal with time series, hierarchies, networks, or multi-variate or textual data, in which questions revolve around finding relationships, clusters, gaps, outliers, and anomalies.

Information visualization problems might be typified by the following examples:

---

[1] The committee thanks Ben Shneiderman of the University of Maryland and Patrick Hanrahan of Stanford University for very helpful inputs to this section.

- Find the strongest correlations in daily stock market performance over 5 years for 10,000 stocks (analyze a set of time series);
- Identify duplicate directories in petabyte-scale hard drives (search through hierarchies);
- Find the most-central nodes in a social network of 500 million users (network analysis); and
- Discover related variables from 100-dimensional data sets with a billion rows (multivariate analysis).

Increasingly, text-analytics projects are searching Web-scale data sets for trending phrases (e.g., Google's culturomics.org), unusual combinations, or anomalous corpora that avoid certain phrases.

There are hundreds of interesting visualization techniques, including simple bar charts, line charts, treemaps, graphs, geographic maps, and textual representations such as tag clouds. Information visualization tools, however, often rely on rich interactions between multiple simultaneous visualizations. For example, a user might select a set of markers in one window, and the tool highlights related markers in all windows, so that relationships can be seen. Such a capability might allow a user to select bars on a timeline indicating earthquakes and, from that, automatically highlight markers on a map and show each earthquake in a scattergram organized by intensity versus number of fatalities, color coded by whether the quake was under water or under land areas. Similarly, the movement of any dynamic query slider immediately filters out the related markers in all views. For example, a user could filter out the quakes whose epicenter was deeper than 3 miles to study the impact of deep quakes only.

The general approach to seeking information is to overview first, zoom and filter, and then find details on demand. This simple notion conveys the basic idea of an exploratory process that has been widely applied. More recently, however, attention in the visual analytics community has shifted to process models for exploration. Such models range from simple 4-step approaches that gather information, re-represent it, develop insights, and present results, to elaborate 16-step models and domain-specific approaches for medical histories, financial transactions, or gene expression data.[2] The process models help guide users through steps that include data cleaning (remove errors, duplicates, missing data, etc.), filtering (select appropriate subsets), aggregation (clustering, grouping, hierarchical organization), and recording insights (marking, annotation, grouping).[3]

The steps in such process models have perceptual, cognitive, and domain-specific aspects that lead researchers to consider visual analytics as a

---

[2] See Thomas and Cook (2005) for an overview.
[3] See, e.g., Perer and Shneiderman (2006).

sense-making process, which requires validation by empirical assessments. While some aspects of interface design and usage can be tested in controlled empirical user studies and expert reviews, information visualization researchers have creatively found new evaluation methods. Often, case studies of usage by actual researchers working with their own data over periods of weeks or months have been used to validate the utility of information visualization tools and visual analytics processes (Shneiderman and Plaisant, 2006).

Crowdsourced Data Acquisition

The idea of coordinating groups of people to perform computational tasks has a long history. Small groups of people were used to catalog scientific observations as early as the 1700s, and groups of hundreds of people were organized to compute and compile tables of mathematical functions in the early part of the 20th century (Grier, 2005). Recently, as the Internet has enabled large-scale organization and interaction of people, there has been a resurgence of interest in crowd-based data gathering and computation. The term "crowdsourcing" is used to describe a number of different approaches, which can be grouped into two general classes: those that leverage human activity, and those that leverage human intelligence. In the former case, data are produced and gathered, and work is performed as a by-product of individuals' behavior on the Web or in other networked environments. In the latter case, groups of people are organized and explicitly tasked with performing a job, solving a problem, or contributing content or other work product.

The first category of crowdsourcing consists of techniques for garnering useful information generated as a by-product of human activity. Such information is sometimes referred to as "data exhaust." For example, search companies can continuously improve their spell checking and recommendation systems using data generated as users enter misspelled search terms and then click on a differently spelled (correct) result. Many Web companies engage in similar types of activity mining, for example, to choose which content or advertising to display to specific users based on search history, access history, demographics, etc. Many other online activities, such as recommending a restaurant, "re-tweeting" an article, and so on, also provide valuable information. The implicit knowledge that is gleaned from user behaviors can be used to create a new predictive model or to augment and improve an existing one.

In these examples, users' online activity is used to predict their intent or discern their interests. Online activity can also be used to understand events and trends in the real world. For example, there has been significant interest lately in continuously monitoring online forums and social media to detect emerging news stories. As another example, Google researchers have demonstrated the ability to accurately detect flu outbreaks in particular geographic regions by noting patterns in search requests about flu symptoms and remedies.[4] Importantly, they demonstrated that such methods sped up the detection of outbreaks by weeks compared to the traditional reporting methods currently used by the Centers for Disease Control and Prevention and others.

Another form of crowdsourced data acquisition is known as participatory sensing (Estrin, 2010). The convergence of sensing, communication, and computational power on mobile devices such as cellular phones creates an unprecedented opportunity for crowdsourcing data. Smartphones are increasingly integrating sensor suites (with data from the Global Positioning System, accelerometers, magnetometers, light sensors, cameras, and so on), and they are capable of processing the geolocalized data and of transmitting them. As such, participatory sensing has become a paradigm for gathering data at global scales, which can reveal patterns of humans in the built environment. Early successes have been in the area of traffic monitoring and congestion prediction,[5] but it is possible to build many applications that integrate physical monitoring with maps. Examples of other applications include monitoring of environmental factors such as air quality, sound pollution, ground shaking (i.e., earthquake detection), and water quality and motion. Urban planning can be aided by the monitoring of vehicular as well as pedestrian traffic. Privacy concerns must be taken into account and handled carefully in some of these cases.

In all the cases described above, data are collected as a by-product of peoples' online or on-network behavior. Another class of crowdsourcing approaches more actively designs online activity with the express purpose of enticing people to provide useful data and processing. "Games with a purpose" are online games that entice users to perform useful work while playing online games (see, e.g., von Ahn and Dabbish, 2008). An early example was the ESP Game, developed at Carnegie Mellon University, in which players listed terms that describe images, simultaneously earning points in the game and labeling the images to aid in future image search queries.

A related approach, called re-captcha,[6] leverages human activity to augment optical character recognition (OCR). In re-captcha, users attempting to access an online resource are presented with two sets of characters to transcribe. One set of characters is known to the algorithm and is presented

---

[4] E.g., Explore Flu Trends Around the World, available at http://www.google.org/flutrends.

[5] E.g., Mobile Millennium, University of California, Berkeley, Snapshot of Mobile Millennium Traffic in San Francisco and the Bay Area, available at http://traffic.berkeley.edu/.

[6] The ReCAPTCHA website is available at http://www.google.com/recaptcha.

in a format that is difficult for machines to identify. The other set of characters presented to the user is a portion of text that an OCR algorithm was unable to recognize. The idea is that by correctly entering the first set of characters, a user verifies that he or she is not a machine, and by entering the second set of characters, the user then effectively performs an OCR task that an OCR algorithm was unable to perform.

HYBRID HUMAN/COMPUTER DATA ANALYSIS

In the crowdsourcing techniques described in the previous section, human input was obtained primarily as part of the data-collection process. More recently, a number of systems have been developed that more explicitly involve people in computational tasks. Although the fields of artificial intelligence and machine learning have made great progress in recent years in solving many problems that were long considered to require human intelligence—for example, natural language processing, language translation, chess playing, winning the television game show Jeopardy, and various prediction and planning tasks—there are still many tasks where human perception, and peoples' ability to disambiguate, understand context, and make subjective judgments, exceed the capabilities of even the most sophisticated computing systems. For such problems, substantial benefit can be obtained by leveraging human intelligence.

While Quinn and Bederson (2011) distinguish human computation from crowdsourcing, defining the former as replacing computers with humans and the latter as "replacing traditional human workers with members of the public," many in both the research community and the general public do not make such a distinction. Thus, crowdsourcing is often used to refer to either type of human involvement, and that convention is followed here.

Some types of crowdsourced systems that can be used to involve people in the process of analyzing data are the following:

- *User-generated content sites.* Wikipedia is a prominent example of a user-generated content site where people create, modify, and update pages of information about a huge range of topics. More specialized sites exist for reviews and recommendations of movies, restaurants, products, and so on. In addition to creating basic content, in many of these systems users are also able to edit and curate the data, resulting in collections of data that can be useful in many analytics tasks.
- *Task platforms.* Much of the interest around crowdsourcing has been focused on an emerging set of systems known as microtask platforms. A microtask platform creates a marketplace in which requesters offer tasks and workers accept and perform the tasks.

Microtasks usually do not require any special training and typically take no longer than 1 minute to complete, although they can take longer. Typical microtasks include labeling images, cleaning and verifying data, locating missing information, and performing subjective or context-based comparisons. One of the leading platforms at present is Amazon Mechanical Turk (AMT). In AMT, workers from anywhere in the world can participate, and there are thought to be hundreds of thousands of people who perform jobs on the system.

Other task-oriented platforms have been developed or proposed to do more sophisticated work. For example, specialized platforms have been developed to crowdsource creative work such as designing logos (e.g., 99designs) or writing code (e.g., TopCoder). In addition, some groups have developed programming languages to encode more sophisticated multistep tasks, such as Turkit (Little et al., 2010), or market-based mechanisms for organizing larger tasks (Shahaf and Horvitz, 2010). These types of platforms can be used to get human participation on a range of analytics tasks, from simple disambiguation to more sophisticated iterative processing.

- *Crowdsourced query processing.* Recently, a number of research efforts have investigated the integration of crowdsourcing with query processing as performed by relational database systems. Traditional database systems are limited in their ability to tolerate inconsistent or missing information, which has restricted the domains in which they can be applied largely to those with structured, fairly clean information. Crowdsourcing based on application programming interfaces (APIs) provides an opportunity to engage humans to help with those tasks that are not sufficiently handled by database systems today. CrowdDB (Franklin et al., 2011) and Qurk (Marcus et al., 2011) are examples of such experimental systems.

- *Question-answering systems.* Question-answering systems are another type of system for enlisting human intelligence. Many different kinds of human-powered or human-assisted sites have been developed. These include general knowledge sites where humans help answer questions (e.g., Cha Cha), general expertise-based sites, where people with expertise in particular topics answer questions on those topics (e.g., Quora), and specialized sites focused on a particular topic (e.g., StackOverflow for computer-programming-related questions).

- *Massive multi-player online games.* Another type of crowdsourcing site uses gamification to encourage people to contribute to solving a problem. Such games can be useful for simulating complex social systems, predicting events (e.g., prediction markets), or for solving specific types of problems. One successful example of the latter type of system is the FoldIt site,[7] where people compete to most accurately predict the way that certain proteins will fold. FoldIt has been competitive with, and in some cases even beaten, the best algorithms for protein folding, even though many of the people participating are not experts.

- *Specialized platforms.* Some crowdsourcing systems have been developed and deployed to solve specialized types of problems. One example is Ushahidi,[8] which provides geographic-based information and visualizations for crisis response and other applications. Another such system is Galaxy Zoo,[9] which enables people to help identify interesting objects in astronomical images. Galaxy Zoo learns the skill sets of its participants over time and uses this knowledge to route particular images to the people who are most likely to accurately detect the phenomena in those images.
- *Collaborative analysis.* This class of systems consists of the crowdsourcing platforms that are perhaps the most directly related to data analytics at present. Such systems enable groups of people to share and discuss data and visualizations in order to detect and understand trends and anomalies. Such systems typically include a social component in which participants can directly engage each other. Examples of such systems include ManyEyes, Swivel, and Sense.us.

As can be seen from the above list, there is currently a tremendous amount of interest in and innovation around crowdsourcing in many forms. In some cases (e.g., crowdsourced query processing and collaborative analysis) crowd resources are being directly used to help make sense of data. In other cases, there is simply the potential for doing so. The next section outlines opportunities and challenges for developing hybrid human/computer analytics systems, as well as the two other areas of human interaction with data discussed above.

OPPORTUNITIES, CHALLENGES, AND DIRECTIONS

Data Visualization and Exploration

Many of the current challenges in visualization and exploration stem from scalability issues. As the volume of data to be analyzed continues to increase, it becomes increasingly difficult to provide useful visual represen-

---

[7] The FoldIt website is available at http://fold.it.
[8] The Ushahidi website is available at http://ushahidi.com.
[9] The Galaxy Zoo website is available at http://www.galaxyzoo.org.

tations and interactive performance for massive data sets. These concerns are not unrelated: interactive analysis is qualitatively different from off-line approaches, particularly when exploration is required.

Aggregation strategies and visual representations are gaining importance as research topics (Shneiderman, 2008; Elmqvist and Fekete, 2010). This is especially true for network visualization, in which billion-node communications or citation graphs are common and petabyte-per-day growth is a reality (Elmqvist et al., 2008; Archambault et al., 2011).

In terms of performance, one would expect that the significant continuing changes in hardware architectures provide an opportunity to address the scalability issue. One appealing research direction is to support massive information visualization by way of specialized hardware. Graphics processing units (GPUs) have become low-cost and pervasive for showing three-dimensional graphics, while other emerging technologies such as data parallel computation platforms and cloud computing infrastructures must also be exploited.

A second area that requires attention is the integration of visualization with statistical methods and other analytic techniques in order to support discovery and analysis. Here, the best strategy appears to lie in combining statistical methods with information visualization (Perer and Shneiderman, 2009). Users can view initial displays of data to gain provisional insights about the distributions, identify errors or missing data, select interesting outliers or clusters, and explore high and low values. At every point they can apply statistical treatments to produce new intermediate data sets, record their insights, select groups for later analysis, or forward promising partial results to colleagues. Often users will need to combine data from several sources and apply domain knowledge to interpret the meaning of a statistical result and visual display. Although the products of an analysis may be compelling displays, the key outcome is insight about the data.

An additional requirement that arises from the interactive nature of many data analysis processes is the need for the analytics system to provide human-understandable feedback to explain analytics results and the steps taken to obtain them. For example, sometimes automated systems produce models that are difficult to understand. Currently, the understandability of the analytical processes is the biggest impediment to using such techniques in decision-making. No CEO is going to make a risky decision using a model they do not understand. Consumers also have problems when automated systems present data they do not understand. Embedding data in a semantic substrate and allowing people to ask high-level questions using interactive tools is an effective way to improve confidence in and utility of an analytics system.

A final area of opportunity is support for group-based analytics. Complex decisions are often made by groups rather than individuals. As data become more complex, support for groups and shared expertise becomes even more important. Visual analytics researchers emphasize the social processes around information visualization, in which teams of 10 to 5,000 analysts may be working on a single problem, such as pharmaceutical drug discovery, oil/gas exploration, manufacturing process control, or intelligence analysis. These teams must coordinate their efforts over weeks or months, generate many intermediate data sets, and combine their insights to support important decisions for corporations or government agencies.

Crowdsourced Data Acquisition and Hybrid Human/Computer Data Analysis

The other two ways that people can participate in the analytics process are by helping to acquire data and by adding human intelligence where existing algorithms and systems technology cannot provide an adequate answer. These two topics are combined because they share many open issues.

One of the main research problems for crowdsourcing is the need to understand, evaluate, and improve the quality of answers obtained from people. Answers from the crowd can be subject to statistical bias, malicious or simply greedy intent (particularly when work is done for pay), or simply incorrect answers due to a lack of expertise. Such problems are exacerbated in some crowdsourcing systems where workers are more or less anonymous and, hence, not fully accountable, and in environments where monetary incentives are used, which can lead to contributors providing large numbers of random or simply incorrect answers. While many traditional statistical tests and error adjustment techniques can be brought to bear on the problem, the environment of crowdsourced work provides new challenges that must be addressed.

Another important area requiring significant work is the design of incentive mechanisms to improve the quality, cost, and timeliness of crowdsourced contributions. Incentive structures currently used include money, status, altruism, and other rewards. Also, because many crowdsourcing platforms are truly global markets, there are concerns about minimum wages, quality of work offered, and potential exploitation of workers that must be addressed.

Participatory sensing provides another set of research challenges. Large-scale sensing deployments can create massive streams of real-time data. These data can be error-prone and context sensitive. Privacy issues must also be taken into account if the sensing is being done based on monitoring individuals' activities. Finally, the sheer volume of data collected can stress even the most advanced computing platforms, particularly if data are to be maintained over long time periods.

An interesting and important problem is that of determining what types of problems are amenable to human solution as opposed to computer solution. This question is related to the question of "AI Completeness" as described by (Shahaf and Amir, 2007). It also leads to what is likely the most important area of future work regarding crowdsourcing and analytics, namely, the design and development of hybrid human/computer systems that solve problems that are too hard for computers or people to solve alone. Designing such systems requires a deep understanding of the relative strengths and weaknesses of human and machine computation.

Given the scale of massive data, it makes sense to try to use computers wherever possible, because people are inherently slower for large number-crunching tasks and their abilities are less scalable. Thus statistical methods and machine-learning algorithms should be used when they can produce answers with sufficient confidence within time and budget constraints. People can be brought to bear to handle cases that need additional clarification or insight. Furthermore, human input can be used to train, validate, and improve models. In the longer term, the expectation is that machines and algorithms will continue to improve in terms of the scale and complexity of the tasks they can accomplish. However, it is likely also that this improvement will lead to an increase in the scope, complexity, and diversity of the analysis questions to be answered. Thus, while the roles and responsibilities of the algorithmic and human components will shift over time, the need to design and develop effective hybrid systems will remain.

REFERENCES

Archambault, D., T. Munzner, and D. Auber. 2011. Tugging graphs faster: Efficiently modifying path-preserving hierarchies for browsing paths. *IEEE Transactions on Visualization and Computer Graphics* 17(3):276-289.

Elmqvist, N., and J.-D. Fekete. 2010. Hierarchical aggregation for information visualization: Overview, techniques, and design guidelines. *IEEE Transactions on Visualization and Computer Graphics* 16(3):439-454.

Elmqvist, N., D. Thanh-Nghi, H. Goodell, N. Henry, and J.-D. Fekete. 2008. ZAME: Interactive Large-Scale Graph Visualization. Pp. 215-222 in *Proceedings of the IEEE Pacific Visualization Symposium 2008* (PacificVIS '08), doi:10.1109/PACIFICVIS.2008.4475479.

Estrin, D. 2010. Participatory sensing: Applications and architecture. *IEEE Internet Computing* 14(1):12-42.

Franklin, M.J., D. Kossmann, T. Kraska, S. Ramesh, and R. Xin. 2011. CrowdDB: Answering queries with crowdsourcing. Pp. 61-72 in *Proceedings of the ACM SIGMOD International Conference on Management of Data, SIGMOD 2011*. Association for Computing Machinery, New York, N.Y.

Grier, D. 2005. *When Computers Were Human*. Princeton University Press, Princeton, N.J.

Little, G., L.B. Chilton, M. Goldman, and R.C. Miller. 2010. TurKit: Human computation algorithms on Mechanical Turk. Pp. 57-66 in *Proceedings of the 23nd Annual ACM Symposium on User Interface Software and Technology*. Association for Computing Machinery, New York, N.Y.

Marcus, A., E. Wu, S. Madden, and R.C. Miller. 2011. Crowdsourced databases: Query processing with people. Pp. 211-214 in *Proceedings of the 2011 Conference on Innovative Data Systems Research (CIDR)*. Available at http://www.cidrdb.org/cidr2011/program. html, pp. 211-214.

Perer, A., and B. Shneiderman. 2006. Balancing systematic and flexible exploration of social networks. *IEEE Transactions on Visualization and Computer Graphics* 12(5):693-700.

Perer, A., and B. Shneiderman. 2009. The importance of integrating statistics and visualization: Long-term case studies supporting exploratory data analysis of social networks. *IEEE Computer Graphics and Applications* 29(3):39-51.

Quinn, A.J., and B.B. Bederson. 2011. Human computation: A survey and taxonomy of a growing field. Pp. 1403-1412 in *Proceedings of the 2011 SIGCHI Conference on Human Factors in Computing*. Association for Computing Machinery, New York, N.Y.

Shahaf, D., and E. Amir. 2007. Towards a theory of AI completeness. Presented at COMMONSENSE 2007: 8th International Symposium on Logical Formalizations of Commonsense Reasoning. In *Logical Formalizations of Commonsense Reasoning: Papers from the AAAI Spring Symposium*. AAAI Technical Report SS-07-05. AAAI Press, Menlo Park, Calif. Available at http://www.ucl.ac.uk/commonsense07/papers/.

Shahaf, D., and E. Horvitz. 2010. Generalized task markets for human and machine computation. Pp. 986-993 in *Proceedings of the 24th AAAI Conference on Artificial Intelligence*. Association for the Advancement of Artificial Intelligence, Palo Alto, Calif.

Shneiderman, B. 2008. Extreme visualization: Squeezing a billion records into a million pixels. Pp. 3-12 in *Proceedings of the ACM SIGMOD 2008 International Conference on the Management of Data*. Association for Computing Machinery, New York, N.Y. Available at http://portal.acm.org/citation.cfm?doid=1376616.1376618.

Shneiderman, B., and C. Plaisant. 2006. Strategies for evaluating information visualization tools: Multi-dimensional in-depth long-term case studies. Pp. 1-7 in *Proceedings of the 2006 Advanced Visual Interfaces Workshop on Beyond Time and Errors: Novel Evaluation Methods for Information Visualization*. Association for Computing Machinery, New York, N.Y.

Thomas, J.J., and K.A. Cook, eds. 2005. *Illuminating the Path: Research and Development Agenda for Visual Analytics*. IEEE Press. Available at http://nvac.pnl.gov/agenda.stm.

von Ahn, L., and L. Dabbish. 2008. Games with a purpose. *Communications of the ACM* 51(8):58-67.

10

The Seven Computational Giants of Massive Data Analysis

One of the major challenges in massive data analysis is that of specifying an overall system architecture. Massive data analysis systems should make effective use of existing (distributed and parallel) hardware platforms; accommodate a wide range of data formats, models, loss functions, and methods; provide an expressive but simple language in which humans can specify their data analysis goals; hide inessential aspects of data analysis from human users while providing useful visualizations of essential aspects of the analysis; provide seamless interfaces to other computational platforms, including scientific measurement platforms and databases; and provide many of capabilities familiar from large-scale databases, such as checkpointing and provenance tracking. These systems should permit appropriate blends of autonomy and human oversight.

Clearly we are far from possessing the ability to design and build such systems. In taking steps toward formulating design principles for massive data analysis systems, the committee notes that a major missing ingredient appears to be agreement on a notion of "middleware," which would connect high-level analysis goals to implementations at the level of hardware and software platforms. The general goal of such middleware is to provide a notion of "reuse," whereby a relatively small set of computational modules can be optimized and exploited in a wide variety of algorithms and analyses.

The field of high-performance computing has faced a similar set of challenges. In that field, a useful step forward was provided by the Berkeley "dwarfs" paper (Asanovic et al., 2006), which specified a set of common problem classes that could help in the design of software for novel supercomputing architectures. In that paper, a "dwarf" represents a computational task that is commonly used and built from a consistent set of fundamental computations and data movements.

This chapter represents a first attempt to define an analogous set of computational tasks for massive data analysis, essentially aiming to provide a taxonomy of tasks that have proved to be useful in data analysis and grouping them roughly according to mathematical structure and computational strategy. Given the vast scope of the problem of data analysis, and the lack of existing general-purpose computational systems for massive data analysis from which to generalize, the committee does not expect this taxonomy to survive a detailed critique. Indeed, the presentation here is not intended to be taken as definitive; rather, the committee wishes it will serve to open a discussion.

Because the subject is massive data, the term "giants" will be used rather than "dwarfs." The committee proposes the following seven giants of statistical data analysis:

1. Basic statistics,
2. Generalized N-body problem,
3. Graph-theoretic computations,
4. Linear algebraic computations,
5. Optimization,
6. Integration, and
7. Alignment problems.

For each of these computational classes, there are computational constraints that arise within any particular problem domain that help to determine the specialized algorithmic strategy to be employed. Such collections of constraints are referred to here as "settings." Important settings for which algorithms are needed include the following:

A. The "default setting" of a single processor with the entire data set fitting in random access memory (RAM);
B. The streaming setting, in which data arrive in quick succession and only a subset can be stored;
C. The disk-based setting, in which the data are too large to store in RAM but fit on one machine's disk;
D. The distributed setting, in which the data are distributed over multiple machines' RAMs or disks; and
E. The multi-threaded setting, in which the data lie on one machine having multiple processors which share RAM.

Most work to date focuses on setting A.

The relative importance of each of these settings is dictated by the state of computer technology and its economics. For example, the relative latencies of the entire hardware stack (e.g., of network communication, disk accesses, RAM accesses, cache accesses, etc.) shifts over time, ultimately affecting the best choice of algorithm for a problem. What constitutes a "fast" or "good" algorithmic solution for a certain giant in a particular setting is determined by the resources (e.g., time, memory, disk space, and so on) with which one is typically concerned. Different algorithms may be designed to achieve efficiency in terms of different resources.

The giants are sometimes associated with, or even defined by, certain conceptual data representations, such as matrices, graphs, sequences, and so on. This is discussed further in Chapter 6. Sitting one conceptual level below such representations are data structures such as arrays, priority queues, hash tables, etc., which are designed to make certain basic operations on the representations efficient. These are discussed as needed in connection with the algorithmic strategies discussed in the remainder of this chapter.

Another way of characterizing the major problems of massive data analysis is to look at the major inferential challenges that must be addressed. These are discussed in earlier chapters, and coincidentally there are also seven of these "inferential giants":

- Assessment of sampling biases,
- Inference about tails,
- Resampling inference,
- Change point detection,
- Reproducibility of analyses,
- Causal inference for observational data, and
- Efficient inference for temporal streams.

The committee has not attempted to map these statistical problems against the algorithmic "giants" discussed next. The algorithmic groupings below are natural when contemplating how to accomplish various analyses within certain computational settings. But the list of inferential giants is a helpful reminder of the ultimate goal of knowledge discovery.

BASIC STATISTICS

Types of Data Analyses and Computations

This class includes basic statistical tasks. Examples include computing the mean, variance, and other moments; estimating the number of distinct elements in a data set; counting the number of elements and finding frequently occurring ones; and calculating order statistics such as the median.

These tasks typically require $O(N)$ calculations for N data points. Some other calculations that arguably fall into this class include sorting and basic forms of clustering.

Such simple statistical computations are widely used in and of themselves, but they also appear inside myriad more complex analyses. For example, multidimensional counts are important in count-based methods such as association rules and in probabilistic inference in graphical models with discrete variables.

Challenges and Examples of Notable Approaches

The problems in this class become more difficult in sublinear models of computation, such as in the streaming model. For many important tasks, identifying the best algorithm is still a subject of ongoing research. For example, an optimal space-efficient approximation algorithm for estimating the number of distinct elements (a problem studied since Flajolet and Martin, 1985) has been discovered only very recently (Kane et al., 2010). Many other problems remain open, notably those involving high-dimensional data. Various data structures can be employed for discrete counts; for an example, see Anderson and Moore (1998). Various "sketching" and "compressed sensing" approaches based on random projections and other forms of subsampling have been developed. These can provide probabilistic accuracy guarantees in exchange for speedup. Examples include sparse (Charikar et al., 2002; Cormode and Muthukrishnan, 2004; Gilbert and Indyk, 2010), dense (Candès et al., 2006; Donoho, 2006), and nonlinear (Agarwal et al., 2005) approaches.

GENERALIZED N-BODY PROBLEMS

Types of Data Analyses and Computations

"Generalized N-body problems" (Gray and Moore, 2001) include virtually any problem involving distances, kernels, or other similarities between (all or many) pairs (or higher-order n-tuples) of points. Such problems are typically of computational complexity $O(N^2)$ or $O(N^3)$ if approached naively. Range searches of various flavors, including spherical and rectangular range searches, are basic multidimensional queries of general use. Nearest-neighbor search problems of various flavors, including all-nearest-neighbors (nearest-neighbor for many query points) and the nearest-neighbor classification problem (which can be performed without a full search), appear in nearest-neighbor classification as well as in modern methods such as nonlinear dimension reduction methods, which are also known as manifold learning methods. Kernel summations appear in both kernel estimators—such as kernel density estimation, kernel regression methods, radial basis function neural networks, and mean-shift tracking—and modern kernelized methods such as support vector machines and kernel principal components analysis (PCA). The kernel summation decision problem (computing the greater of two kernel summations) occurs in kernel discriminant analysis as well as in support vector machines. Other instances of this type of computation include k-means, mixtures of Gaussians clustering, hierarchical clustering, spatial statistics of various kinds, spatial joins, the Hausdorff set distance, and many others.

Challenges and Examples of Notable Approaches

A main challenge for such problems lies in dealing with high-dimensional data, because the bounds used in typical algorithmic approaches become less effective. For example, designing nearest-neighbor algorithms that do not suffer from the curse of dimensionality is an important topic worthy of further study. Although some progress in this direction has been made (e.g., Freeman, 2011), more research is needed. Various non-Euclidean metrics—such as edit distances used in computational biology, earth-mover distances used in computer vision, and others—appear to pose even greater difficulties. For exact computations, some approaches that can be highly effective include multidimensional data structures such as kd-trees and cover-trees (Beygelzimer et al., 2006). Such algorithms can achieve $O(N \log N)$ run times for all-pairs problems, which would naively require $O(N^2)$ operations (Ram et al., 2009a). Kernel summations can be accurately approximated using sophisticated hierarchical series approximation methods within trees (Lee and Gray, 2006). To achieve greater efficiency in higher dimensionalities, but at the cost of approximation guarantees holding only with high probability, random projections (sampling the dimensions) can be employed (Andoni and Indyk, 2008). Related sketching ideas such as in Church et al. (2006) can be used for distance computations. Sampling ideas can also be powerfully employed within tree-based algorithms for increased accuracies at the cost of greater preprocessing time (Ram et al., 2009b; Lee and Gray, 2009).

GRAPH-THEORETIC COMPUTATIONS

Types of Data Analyses and Computations

This class includes problems that involve traversing a graph. In some cases the graph is the data, and in other cases the statistical model takes the form of a graph, as in the case of graphical models. Common statistical computations that are employed on (data) graphs include betweenness, centrality, and commute distances; these are used to identify nodes or communities of interest. Despite the simple definition of such statistics, major computational challenges arise in large-scale, sparse graphs. When the statistical model takes the form of a graph, graph-search algorithms continue to remain important, but there is also a need to compute marginal probabilities and conditional probabilities over graphs, operations generally referred to as "inference" in the graphical models literature.

In models with all discrete variables, graphical model inference requires deeply nested (many-variable) summations. In models with all Gaussian variables, inference becomes a linear algebra problem (thus becoming a member of the next giant instead). Many graph computations can in fact be posed as linear algebra problems. Other sorts of graph-theoretic computations occur in manifold learning methods. For example, the Isomap method requires an all-pairs-shortest-paths computation. Another example is single-linkage hierarchical clustering, which is equivalent to computing a minimum spanning tree. Note that both of these are examples of Euclidean graph problems, which actually become distance-based, or N-body-type problems (the previous giant).

Challenges and Examples of Notable Approaches

The challenge regime of graph theoretic problems is that of graphs with high interconnectivity in general. For example, in graphical model inference problems, the challenging regime consists of graphs with large maximal clique size. Fundamental graph computations, such as shortest-path calculations, can pose significant challenges for graphs that do not fit in RAM (e.g., due to the latency of remote memory access). Notable recent approaches include sampling (Sarkar et al., 2008) and disk-based (Ajwani et al., 2006; Sarkar and Moore, 2010) ideas.

Comprehensive parallel/distributed approaches to computing graph primitives can be built from work in sparse linear algebra (Gilbert et al., 2007; Kang et al., 2009), or they can use graph concepts more directly (Madduri et al., 2007). Powerful links between graph partitioning and linear algebraic reconditioning have been employed (Spielman and Teng, 2004; Andersen et al., 2006; Leskovec et al., 2009). Notable approaches for graphical model inferences include transformation of the problem from one of summation to one of optimization via variational methods (Jordan et al., 1999; Wainwright and Jordan, 2003), sampling approaches (Dillon and Lebanon, 2010), and parallel/distributed approaches (Gonzalez et al., 2009).

LINEAR ALGEBRAIC COMPUTATIONS

Types of Data Analyses and Computations

This class includes all the standard problems of computational linear algebra, including linear systems, eigenvalue problems, and inverses. A large number of linear models, including linear regression, PCA, and their many variants, result in linear algebraic computations. Many of these are well-solved by generic linear algebra approaches. There are at least two important differentiators, however. One is the fact that, in statistical problems, the optimization in learning—this is what the eigendecomposition of PCA is doing, optimizing a linear convex training error—need not necessarily be performed to high accuracy. This is because one wants to optimize generalization error and not training error, and thus optimizing the training error to high accuracy may be overfitting. Another difference is that multivariate statistics arguably has its own matrix form, that of a kernel (or Gram) matrix. This is significant because much of computational linear algebra involves techniques specialized to take advantage of certain matrix structures. In kernel methods such as Gaussian process regression (kriging) or kernel PCA, the kernel matrix can be so large as to prohibit even storing the matrix explicitly, motivating matrix-free algorithms if possible.

Challenges and Examples of Notable Approaches

Matrices that do not exhibit quickly decaying spectra—a feature that indicates strong structure that can be exploited computationally—and near-singular matrices represent two common challenges for linear algebraic problems. Many problems (such as PCA and linear regression) appear to be harder once the L_2 norm is replaced by other L_p norms, notably the L_1 norm. Probabilistic relaxations of the problem can be employed, sampling from the rows and/or columns of the matrix (Frieze et al., 1998; Drineas et al., 2004). As in generalized N-body problems, the use of sampling within tree data structures can provide increased accuracies (Holmes et al., 2009). For kernel matrix computations as performed in Gaussian process regression, some effective approaches include greedy subsection selection (Smola and Bartlett, 2001; Ouimet and Bengio, 2005; Bo and Sminchisescu, 2008), conjugate gradient-type methods requiring only the ability to multiply the kernel matrix by a vector (changing the core of the computation to a generalized N-body problem; Gray, 2004), and random sampling to compute the kernel (Rahimi and Recht, 2008).

OPTIMIZATIONS

Types of Data Analyses and Computations

Within massive data, all the standard subclasses of optimization problems appear, from unconstrained to constrained, both convex and non-convex. Linear algebraic computations are arguably a special case of optimization problems, and they are certainly the main subroutine of (second-order) optimization algorithms. Non-trivial optimizations have appeared in statistical methods from early on, and they appear increasingly frequently as methods have become more sophisticated. For example, linear programming (LP), quadratic programming (QP), and second-order cone programming appear in various forms of support vector machines as well as more recent classifiers, and semidefinite programming appears in recent manifold learning methods such as maximum variance unfolding. It is only a matter of time before other standard types of optimization problems, such as geometric programming (which has not yet been adopted within statistical fields), are applied in statistical methods. As touched on above for linear algebra computations, optimization in statistical problems is often focused on empirical loss functions, which are proxies for expectations. This is usefully formalized by the field of stochastic approximation.

Challenges and Examples of Notable Approaches

Optimization problems (mathematical programs) with a very large number of variables and/or constraints represent a primary difficulty in optimization. The global solution of non-convex problems remains an open problem. Problems with integer constraints come up against a fundamental hardness in computer science, that of combinatorial problems. In general, exploiting the particular mathematical forms of certain optimization problems can lead to more effective optimizers, as in the case of support vector machine-like quadratic programs (Platt, 1998). Some mathematical forms have special properties, for example, submodularity (Chechetka and Guestrin, 2007), which can provably be exploited for greater efficiencies. Distributed optimization, both centralized and asynchronous (Tsitsiklis et al., 1986; Nedic and Ozdaglar, 2009; Boyd et al., 2011; Duchi et al., 2010), is of increasing interest, and creating effective algorithms remains a challenge. A powerful technique of stochastic programming, called stochastic approximation or online learning, can be regarded as a random sampling approach to gradient descent, and this has been effective for many methods, including linear classifiers (Zhang, 2004; Shalev-Shwartz et al., 2007; Bottou and Bousquet, 2008; Nemirovski et al., 2009; Ouyang and Gray, 2010). The parallelization of online approaches remains relatively open.

INTEGRATION

Types of Data Analyses and Computations

Integration of functions is a key class of computations within massive data analysis. Integrations are needed for fully Bayesian inference using any model, and they also arise in non-Bayesian statistical settings, most notably random effects models. The integrals that appear in statistics are often expectations, and thus they have a special form.

Challenges and Examples of Notable Approaches

The frontier of capability in integration surrounds high-dimensional integrals, as low-dimensional integrals are generally well-treated by quadrature methods. The kinds of integrals arising in Bayesian statistical models are typically of high dimension for modern problems. The default approach for high-dimensional integration is Markov Chain Monte Carlo (MCMC; Andrie et al., 2003). In the case of certain sequential models, the approach becomes that of sequential Monte Carlo, which is also known as particle filtering (Doucet et al., 2001).

Effective alternatives to MCMC include approximate Bayesian computation (ABC) methods, which operate on summary data (such as population means or variances) to achieve accelerations in some cases (Marjoram et al., 2003), and population Monte Carlo, a form of adaptive importance sampling (Capp et al., 2004). Alternative approaches based on low-discrepancy sets improve over Monte Carlo integration in some cases (Paskov and Traub, 1995). Variational methods provide a general way to convert integration problems into optimization problems (Wainwright and Jordan, 2003).

Because of the inherent difficulty of high-dimensional integration, a common strategy is to change the inference formulation away from a full Bayesian one—for example, in maximum a posteriori inference and empirical Bayesian inference, part of the integration problem is skirted via the use of optimization-based point estimation.

ALIGNMENT PROBLEMS

Types of Data Analyses and Computations

The class of alignment problems consists of various types of problems involving matchings between two or more data objects or data sets, such as the multiple sequence alignments commonly used in computational biology, the matching of catalogs from different instruments in astronomy, the matching of objects between images, and the correspondence between synonymous words in text analysis. Such non-trivial problems are often critical in the data fusion that must often be carried out before further data analyses can be performed.

Challenges and Examples of Notable Approaches

Such problems are often combinatorial in nature, and thus various forms of problem-specific constraints are generally exploited to make the computations efficient. The sequential structure of genomics problems naturally lead to dynamic programming solutions, but for larger scales, greedy hierarchical solutions (Higgins and Sharpe, 1988) and hidden Markov models (Grasso and Lee, 2004) are often used. For the catalog cross-match problem, which has the structure of a generalized N-body problem, spatial locality can be exploited using parallel in-database spatial indexing methods (Gray et al., 2006; Nieto-Santisteban et al., 2006). The problem of finding correspondences between features of objects, such as faces between images, can in principle be treated with a matching algorithm (Bertsekas, 1988), but it must also account for various invariances (Zokai and Wolberg, 2005). The synonymy problem in text can be handled by approaches such as linear dimension-reduction models, taking the form of a linear algebraic problem (Landauer et al., 1998).

DISCUSSION

Two of the most pervasive strategies for achieving computational efficiency are sampling and parallel/distributed computing. Sampling is discussed further in Chapter 8, where the focus is on the statistical aspects rather than the computational and algorithmic aspects touched on here. Current and emerging technologies for parallel and distributed computing are discussed in Chapter 3, where the focus is on architectural issues rather than the algorithmic aspects touched on here.

Looking across all of the seven giants for common themes, the following are evident:

- State-of-the-art algorithms exist that can provide accelerations of major practical importance, by significantly changing the runtime order, for example, from $O(N^2)$ to $O(N \log N)$.
- The "non-default" settings—streaming, disk-based, distributed, multi-threaded—are quite important, yet mostly under-explored in terms of research effort.

- High dimensionality in the number of variables is a persistent challenge to obtaining computational efficiency, and this demands ongoing research effort.
- Most of the best fast algorithms described in this chapter have only been demonstrated in research implementations. More work is required to create robust and reliable software before these algorithms can be used widely in practice.

The combination of the seven giants and the five settings A through E, identified early in this chapter, imply a table of research frontiers. As noted early in this chapter, most work to date focuses on setting A, the "default setting" of a single processor with the entire data set fitting in random access memory, so much work remains in order to more completely explore this space of algorithms and settings.

REFERENCES

Agarwal, P.K., S. Har-Peled, and K.R. Varadarajan. 2005. Geometric approximation via core-sets. Pp. 1-30 in *Combinatorial and Computational Geometry* (J.E. Goodman, J. Pach, and E. Welzl, eds.). Cambridge University Press, New York, N.Y.

Ajwani, D., R. Dementiev, and U. Mayer. 2006. A computational study of external-memory BFS algorithms. Pp. 601-610 in *Proceedings of the 17th ACM-SIAM Symposium on Discrete Algorithms*. Society for Industrial and Applied Mathematics, Philadelphia, Pa.

Andersen, R., F. Chung, and K. Lang. 2006. Local graph partitioning using PageRank vectors. Pp. 475-486 in *Proceedings of the 47th Annual IEEE Symposium on Foundations of Computer Science*. IEEE Computer Society, Washington, D.C. Available at http://ieeexplore.ieee.org/xpl/tocresult.jsp?isnumber=4031330.

Anderson, B., and A. Moore. 1998. Ad-trees for fast counting and for fast learning of association rules. Pp. 134-138 in *Knowledge Discovery from Databases Conference*. KDD-98. Available at http://www.aaai.org/Papers/KDD/1998/KDD98-020.pdf.

Andoni, A., and P. Indyk. 2008. Near-optimal hashing algorithms for approximate nearest neighbor in high dimensions. *Communications of the ACM* 51(1):117-122.

Andrie, C., N. de Freitas, A. Doucet, and M.I. Jordan. 2003. An introduction to MCMC for machine learning. *Machine Learning* 50(1-2):5-43.

Asanovic, K., R. Bodik, B.C. Catanzaro, J.J. Gebis, P. Husbands, K. Keutzer, D.A. Patterson, W.L. Plishker, J. Shalf, S.W. Williams, and K.A. Yelick. 2006. *The Landscape of Parallel Computing Research: A View from Berkeley*. University of California, Berkeley, Technical Report No. UCB/EECS-2006-183. December 18. Available at http://www.eecs.berkeley.edu/Pubs/TechRpts/2006/EECS-2006-183.html.

Bertsekas, D. 1988. The auction algorithm: A distributed relaxation method for the assignment problem. *Annals of Operations Research* 14(1):105-123.

Beygelzimer, A., S. Kakade, and J. Langford. 2006. Cover trees for nearest neighbor. Pp. 97-104 in *Proceedings of the 23rd International Conference on Machine Learning*. Association for Computing Machinery, New York, N.Y.

Bo, L., and C. Sminchisescu. 2008. Greedy block coordinate descent for large scale Gaussian process regression. Pp. 43-52 in *Proceedings of the Twenty-Fourth Conference on Uncertainty in Artificial Intelligence*. Association for Uncertainty in Artificial Intelligence Press, Press, Corvallis, Ore. Available at http://uai.sis.pitt.edu/displayArticles.jsp?mmnu=1&smnu=1&proceeding_id=24.

Bottou, L., and O. Bousquet. 2008. The tradeoffs of large scale learning. In *Advances in Neural Information Processing Systems 20* (J. Platt, D. Koller, Y. Singer, and S. Roweis, eds.). NIPS Foundation. Available at http://books.nips.cc.

Boyd, S., N. Parikh, E. Chu, B. Peleato, and J. Eckstein. 2011. Distributed optimization and statistical learning via the alternating direction method of multipliers. *Foundations and Trends in Machine Learning* 3(1):1-122.

Candès, E.J., J. Romberg, and T. Tao. 2006. Stable signal recovery from incomplete and inaccurate measurements. *Communications on Pure and Applied Mathematics* 59(8): 1208-1223.

Capp, O., A. Guillin, J.-M. Marin, and C.P. Robert. 2004. Population Monte Carlo. *Journal of Computational and Graphical Statistics* 13:907-929.

Charikar, M., K. Chen, and M. Farach-Colton. 2002. Finding frequent items in data streams. Pp. 693-703 in *Proceedings of the 29th International Colloquium on Automata, Languages and Programming*. Springer-Verlag, London.

Chechetka, A., and C. Guestrin. 2008. Efficient principled learning of thin junction trees. In *Advances in Neural Information Processing Systems 20* (J.C. Platt, D. Koller, Y. Singer, and S. Roweis, eds.). NIPS Foundation. Available at http://books.nips.cc.

Church, K.W., P. Li, and T.J. Hastie. 2007. Conditional random sampling: A sketch-based sampling technique for sparse data. In *Advances in Neural Information Processing Systems 19*. NIPS Foundation. Available at http://books.nips.cc.

Cormode, G., and S. Muthukrishnan. 2005. Improved data stream summaries: The count-min sketch and its applications. *Journal of Algorithms* 55(1):58-75.

Dillon, J., and G. Lebanon. 2010. Stochastic composite likelihood. *Journal of Machine Learning Research* 11:2597-2633.

Donoho, D.L. 2006. Compressed sensing. *IEEE Transactions on Information Theory* 52: 1289-1306.

Doucet, A., N. De Freitas, and N. Gordon, eds. 2001. *Sequential Monte Carlo Methods in Practice*. Springer, New York, N.Y.

Drineas, P., R. Kannan, and M.W. Mahoney. 2004. Fast Monte Carlo algorithms for matrices ii: Computing a low-rank approximation to a matrix. *SIAM Journal on Computing* 36:2006.

Duchi, J., A. Agarwal, and M. Wainwright. 2010. Distributed dual averaging in networks. In *Advances in Neural Information Processing Systems 23* (J. Lafferty, C.K.I. Williams, J. Shawe-Taylor, R.S. Zemel, and A. Culotta, eds.). NIPS Foundation. Available at http://books.nips.cc.

Flajolet, P., and G. Martin. 1985. Probabilistic counting algorithms for data base applications. *Journal of Computer and System Sciences* 31(2):182-209.

Freeman, W.T. 2011. Where computer vision needs help from computer science. Pp. 814-819 in *Proceedings of the Twenty-Second Annual ACM-SIAM Symposium on Discrete Algorithms*. Society for Industrial and Applied Mathematics, Philadelphia, Pa.

Frieze, A., R. Kannan, and S. Vempala. 1998. Fast Monte Carlo algorithms for finding low-rank approximations. Pp. 370-378 in *Proceedings of the 39th Annual IEEE Symposium on Foundations of Computer Science*. IEEE Computer Society, Washington, D.C. Available at http://ieeexplore.ieee.org/xpl/aboutJournal.jsp?punumber=5965.

Gilbert, A., and P. Indyk. 2010. Sparse recovery using sparse matrices. *Proceedings of IEEE* 98(6):937-947.

Gilbert, J.R., S. Reinhardt, and V. Shah. 2007. High performance graph algorithms from parallel sparse matrices. Pp. 260-269 in *Proceedings of the 8th International Conference on Applied Parallel Computing: State of the Art in Scientific Computing.* Springer-Verlag, Berlin.

Gonzalez, J., Y. Low, and C. Guestrin. 2009. Residual splash for optimally parallelizing belief propagation. Pp. 177-184 in *Proceedings of the 12th International Conference on Artificial Intelligence and Statistics.* Available at http://jmlr.org/proceedings/papers/v5/.

Gonzalez, J., Y. Low, and C. Guestrin. 2009. Residual splash for optimally parallelizing belief propagation. In *Artificial Intelligence and Statistics (AISTATS),* Clearwater Beach, Fla., April.

Grasso, C., and C. Lee. 2004. Combining partial order alignment and progressive multiple sequence alignment increases alignment speed and scalability to very large alignment problems. *Bioinformatics* 20(10):1546-1556.

Gray, A.G. 2004. *Fast Kernel Matrix-Vector Multiplication with Application to Gaussian Process Regression.* Technical report. Carnegie Mellon University, Pittsburgh, Pa.

Gray, A.G., and A.W. Moore. 2001. N-Body problems in statistical learning. In *Advances in Neural Information Processing Systems 13* (T.K. Leen, T.G. Dietterich, and V. Tresp, eds). NIPS Foundation. Available at http://books.nips.cc.

Gray, J., M.A. Nieto-Santisteban, and A.S. Szalay. 2006. *The Zones Algorithm for Finding Points-Near-a-Point or Cross-Matching Spatial Datasets.* Microsoft Research Technical Report MSR-TR-2006-52. Available at http://research.microsoft.com/apps/pubs/default. aspx?id=64524.

Higgins, D., and P. Sharpe. 1988. Clustal: A package for performing multiple sequence alignment on a microcomputer. *Gene* 73(1):237-244.

Holmes, M., A.G. Gray, and C. Isbell. 2009. QUIC-SVD: Fast SVD using cosine trees. In *Advances in Neural Information Processing Systems 21.* NIPS Foundation. Available at http://books.nips.cc.

Jordan, M.I., Z. Ghahramani, T.S. Jaakkola, and L.K. Saul. 1999. An introduction to variational methods for graphical models. *Machine Learning* 37:183-233.

Kane, D.M., J. Nelson, and D.P. Woodruff. 2010. An optimal algorithm for the distinct elements problem. Pp. 41-52 in *Proceedings of the 29th ACM SIGMOD-SIGACT-SIGART Symposium on Principles of Database Systems.* Association for Computing Machinery, New York, N.Y.

Kang, U., C. Tsourakakis, and C. Faloutsos. 2009. Pegasus: A peta-scale graph mining system—Implementation and observations. Pp. 229-238 in *Proceedings of the 2009 Ninth IEEE International Conference on Data Mining.* IEEE Computer Society, Washington, D.C.

Landauer, T.K., D. Laham, and P. Foltz. 1998. Learning human-like knowledge by singular value decomposition: A progress report. In *Advances in Neural Information Processing Systems 10.* NIPS Foundation. Available at http://books.nips.cc.

Lee, D., and A.G. Gray. 2006. Faster Gaussian summation: Theory and experiment. Pp. 281-288 in *Proceedings of the Twenty-Second Annual Conference on Uncertainty in Artificial Intelligence.* AUAI Press, Arlington, Va. Available at http://uai.sis.pitt.edu/displayArticles. jsp?mmnu=1&smnu=1&proceeding_id=22.

Lee, D., and A.G. Gray. 2009. Fast high-dimensional kernel summations using the Monte Carlo multipole method. In *Advances in Neural Information Processing Systems 21.* NIPS Foundation. Available at http://books.nips.cc.

Leskovec, J., K.J. Lang, A. Dasgupta, and M.W. Mahoney. 2009. Community structure in large networks: Natural cluster sizes and the absence of large well-defined clusters. *Internet Mathematics* 6(1):29-123.

Madduri, K., D.A. Bader, J.W. Berry, J.R. Crobak, and B.A. Hendrickson. 2007. Multithreaded algorithms for processing massive graphs. In *Petascale Computing: Algorithms and Applications* (D. Bader, ed.). Chapman and Hall, CRC Press.

Marjoram, P., J. Molitor, V. Plagnol, and S. Tavar. 2003. Markov chain Monte Carlo without likelihoods. *Proceedings of the National Academy of Sciences U.S.A.* 100(26): 15324-15328.

Nedic, A., and A. Ozdaglar. 2009. Distributed subgradient methods for multi-agent optimization. *IEEE Transactions on Automatic Control* 54(1):48-61.

Nemirovski, A., A. Juditsky, G. Lan, and A. Shapiro. 2009. Robust stochastic approximation approach to stochastic programming. *SIAM Journal of Optimization* 19(4):1574-1609.

Nieto-Santisteban, M.A., A.R. Thakar, A.S. Szalay, and J. Gray. 2006. Large-scale query and XMatch, entering the parallel zone. Pp. 493-496 in *Astronomical Data Analysis Software and Systems XV* (C. Gabriel, C. Arviset, D. Ponz, and S. Enrique, eds.). *Astronomical Society of the Pacific Conference Series*, Volume 351. Available at http://www.aspbooks. org/a/volumes/article_details/?paper_id=3461.

Ouimet, M., and Y. Bengio. 2005. Greedy spectral embedding. Pp. 253-260 in *Proceedings of the 10th International Workshop on Artificial Intelligence and Statistics*. Available at http://www.gatsby.ucl.ac.uk/aistats/AIabst.htm.

Ouyang, H., and A.G. Gray. 2010. Fast stochastic Frank-Wolfe algorithms for nonlinear SVMs. Pp. 245-256 in *Proceedings of the Tenth SIAM International Conference on Data Mining*. Society of Industrial and Applied Mathematics, Philadelphia, Pa.

Paskov, S.H., and J.F. Traub. 1995. Faster valuation of financial derivatives. *Journal of Portfolio Management* 22(1):113-123.

Platt, J. 1998. Fast training of support vector machines using sequential minimal optimization. In *Advances in Kernel Methods—Support Vector Learning* (B. Schoelkopf, C. Burges, and A. Smola, eds.). MIT Press, Cambridge, Mass.

Rahimi, A., and B. Recht. 2008. Random features for large-scale kernel machines. *Advances in Neural Information Processing Systems* 20:1177-1184.

Ram, P., D. Lee, W. March, and A.G. Gray. 2009a. Linear-time algorithms for pairwise statistical problems. In *Advances in Neural Information Processing Systems 22*. NIPS Foundation. Available at http://books.nips.cc.

Ram, P., D. Lee, H. Ouyang, and A.G. Gray. 2009b. Rank-approximate nearest neighbor search: Retaining meaning and speed in high dimensions. In *Advances in Neural Information Processing Systems 22*. NIPS Foundation. Available at http://books.nips.cc.

Sarkar, P., and A. Moore. 2010. Fast nearest-neighbor search in disk-resident graphs. Pp. 513-522 in *Proceedings of the 16th ACM SIGKDD International Conference on Knowledge Discovery and Data Mining*. Association for Computing Machinery, New York, N.Y.

Sarkar, P., A.W. Moore, and A. Prakash. 2008. Fast incremental proximity search in large graphs. Pp. 896-903 in *Proceedings of the 25th International Conference on Machine Learning*. Association for Computing Machinery, New York, N.Y.

Shalev-Shwartz, S., Y. Singer, and N. Srebro. 2007. Pegasos: Primal Estimated sub-GrAdient SOlver for SVM. Pp. 807-814 in *Proceedings of the 24th International Conference on Machine Learning*. Association for Computing Machinery, New York, N.Y.

Smola, A., and P. Bartlett. 2001. Sparse greedy Gaussian process regression. In *Advances in Neural Information Processing Systems 13* (T.K. Leen, T.G. Dietterich, and V. Tresp, eds.). NIPS Foundation. Available at http://books.nips.cc.

Spielman, D.A., and S.-H. Teng. 2004. Nearly linear time algorithms for graph partitioning, graph sparsification, and solving linear systems. Pp. 81-90 in *Proceedings of the 36th Annual ACM Symposium on Theory of Computing (STOC '04)*. Association for Computing Machinery, New York, N.Y.

Tsitsiklis, J.N., D.P. Bertsekas, and M. Athans. 1986. Distributed asynchronous deterministic and stochastic gradient optimization algorithms. *IEEE Transactions on Automatic Control* 31(9):803-812.

Wainwright, M.J., and M.I. Jordan. 2003. Graphical models, exponential families and variational inference. *Foundations and Trends in Machine Learning* 1:1-305.

Zhang, T. 2004. Solving large scale linear prediction problems using stochastic gradient descent algorithms. Pp. 919-926 in *Proceedings of the 21st International Conference on Machine Learning*. Association for Computing Machinery, New York, N.Y.

Zokai, S., and G. Wolberg. 2005. Image registration using log-polar mappings for recovery of large-scale similarity and projective transformations. *IEEE Transactions on Image Processing* 14(10):1422-1434.

11

Conclusions

This report aims to increase the level of awareness of the intellectual and technical issues surrounding the analysis of massive data. This is not the first report written on massive data, and it will not be the last, but given the major attention currently being paid to massive data in science, technology, and government, the committee believes that it is a particularly appropriate time to be considering these issues.

This final section begins by summarizing some of the key conclusions from the report. It then provides a few additional concluding remarks. The study that led to this report reached the following conclusions:

- Recent years have seen rapid growth in parallel and distributed computing systems, developed in large part to serve as the backbone of the modern Internet-based information ecosystem. These systems have fueled search engines, electronic commerce, social networks, and online entertainment, and they provide the platform on which massive data analysis issues have come to the fore. Part of the challenge going forward is the problem of scaling these systems and algorithms to ever-larger collections of data. It is important to acknowledge, however, that the goals of massive data analysis go beyond the computational and representational issues that have been the province of classical search engines and database processing to tackling the challenges of statistical inference, where the goal is to turn data into knowledge and to support effective decision-making. Assertions of knowledge require control over errors, and a major part of the challenge of massive data analysis is that of developing statistically well-founded procedures that provide control over errors in the setting of massive data, recognizing that these procedures are themselves computational procedures that consume resources.

- There are many sources of potential error in massive data analysis, many of which are due to the interest in "long tails" that often accompany the collection of massive data. Events in the "long tail" may be vanishingly rare even in a massive data set. For example, in consumer-facing information technology, where the goal is increasingly that of providing fine-grained, personalized services, there may be little data available for many individuals even in very large data sets. In science, the goal is often that of finding unusual or rare phenomena, and evidence for such phenomena may be weak, particularly when one considers the increase in error rates associated with searching over large classes of hypotheses. Other sources of error that are prevalent in massive data include the high-dimensional nature of many data sets, issues of heterogeneity, biases arising from uncontrolled sampling patterns, and unknown provenance of items in a database. In general, data analysis is based on assumptions, and the assumptions underlying many classical data analysis methods are likely to be broken in massive data sets.

- Massive data analysis is not the province of any one field, but is rather a thoroughly interdisciplinary enterprise. Solutions to massive data problems will require an intimate blending of ideas from computer science and statistics, with essential contributions also needed from applied and pure mathematics, from optimization theory, and from various engineering areas, notably signal processing and information theory. Domain scientists and users of technology also need to be engaged throughout the process of designing systems for massive data analysis. There are also many issues surrounding massive data (most notably privacy issues) that will require input from legal scholars, economists, and other social scientists, although these aspects have not been covered in the current report. In general, by bringing interdisciplinary perspectives to bear on massive data analysis, it will be possible to discuss trade-offs that arise when one jointly considers the computational, statistical, scientific, and human-centric constraints that frame a problem. When considering parts of the problem in isolation, one may end up trying to solve a problem that is more general than is required, and there may be no feasible solution to that broader problem; a suitable cross-disciplinary outlook can point researchers toward an essential refocusing. For example, absent appropriate insight, one might be led to analyzing worst-case algorithmic behavior, which can be very difficult or misleading, whereas a look at the totality of a problem could reveal that average-case algorithmic behavior is quite appropriate from a statistical perspective. Similarly, knowledge of typical query generation might allow one to confine an analysis to a relatively simple subset of all possible queries that would have to be considered in a more general case. And the difficulty of parallel programming in the most general settings may be sidestepped by focusing on useful classes of statistical algorithms that can be implemented with a simplified set of parallel programming motifs; moreover, these motifs may suggest natural patterns of storage and access of data on distributed hardware platforms.

- Although there are many sources of data that are currently fueling the rapid growth in data volume, a few forms of data create particularly interesting challenges. First, much current data involve human language and speech, and increasingly the goal with such data is to extract aspects of the semantic meaning underlying the data. Examples include sentiment analysis, topic models of documents, relational modeling, and the full-blown semantic analyses required by question-answering systems. Second, video and image data are increasingly prevalent, creating a range of challenges in large-scale compression, image processing, computational vision, and semantic analysis. Third, data are increasingly labeled with geo-spatial and temporal tags, creating challenges in maintaining coherence across spatial scales and time. Fourth, many data sets involve networks and graphs, with inferential questions hinging on semantically rich notions such as "centrality" and "influence." The deeper analyses required by data sources such as these involve difficult and unsolved problems in artificial intelligence and the mathematical sciences that go beyond near-term issues of scaling existing algorithms. The committee notes, however, that massive data itself can provide new leverage on such problems, with machine translation of natural language a frequently cited example.

- Massive data analysis creates new challenges at the interface between humans and computers. As just alluded to, many data sets require semantic understanding that is currently beyond the reach of algorithmic approaches and for which human input is needed. This input may be obtained from the data analyst, whose judgment is needed throughout the data analysis process, from the framing of hypotheses to the management of trade-offs (e.g., errors versus time) to the selection of questions to pursue further. It may also be obtained from crowdsourcing, a potentially powerful source of inputs that must be used with care, given the many kinds of errors and biases that can arise. In either case, there are many challenges that need to be faced in the design of effective visualizations and interfaces and, more generally, in linking human judgment with data analysis algorithms.

- Many data sources operate in real time, producing data streams that can overwhelm data analysis pipelines. Moreover, there is often a desire to make decisions rapidly, perhaps also in real time. These temporal issues provide a particularly clear example of the need for further dialog between statistical and computational researchers. Statistical research has rarely considered constraints due to real-time decision-making in the development of data analysis algorithms, and computational research has rarely considered the computational complexity of algorithms for managing statistical risk.

- There is a major need for the development of "middleware"— software components that link high-level data analysis specifications with low-level distributed systems architectures. Chapter 10 attempts to provide an initial set of suggestions in this regard. As discussed there, much of the work on these software components can borrow from tools already developed in scientific computing instances, but the focus will need to change, with algorithmic solutions constrained by statistical needs. There is also a major need for software targeted to end users, such that relatively naive users can carry out massive data analysis without a full understanding of the underlying systems issues and statistical issues. However, this is not to suggest that the end goal of massive data analysis software is to develop turnkey solutions. The exercise of effective human judgment will always be required in data analysis, and this judgment needs to be based on an understanding of statistics and computation. The development of massive data analysis systems needs to proceed in parallel with a major effort to educate students and the workforce in statistical thinking and computational thinking.

The remainder of this chapter provides a few closing remarks on massive data analysis, focusing on issues that have not been highlighted earlier in the report.

The committee is agnostic as to whether a new name, such as "data science," needs to be invoked in discussing research and development in massive data analysis. To the extent that such names invoke an interdisciplinary perspective, the committee feels that they are useful.

In particular, the committee recognizes that industry currently has major needs in the hiring of computer scientists with an appreciation of statistical ideas and statisticians with an appreciation of computational ideas. The use of terms such as "data science" indicates this interdisciplinary hiring profile. Moreover, the existing needs of industry suggest that academia should begin to develop programs that train bachelors- and masters-level students in massive data analysis (in addition to programs at the Ph.D. level). Several such efforts are already under way, and many more are likely to emerge in the next few years. It is perhaps premature to suggest curricula for such programs, particularly given that much of the foundational research in massive data analysis remains to be done. Even if such programs minimally solve the difficult problem of finding room in already-full curricula in computer science and statistics, so that complementary ideas from the other field are taught, they will have made very significant progress.

A broader problem is that training in massive data analysis will require experience with massive data and with computational infrastructure that permits the real problems associated with massive data to be revealed. The availability of benchmarks, repositories (of data and software), and computational infrastructure will be a necessity in training the next generation of "data scientists." The same point, of course, can be made for academic research: significant new ideas will only emerge if academics are exposed to real-world massive data problems.

The committee emphasizes that massive data analysis is not one problem or one methodology. Data are often heterogeneous, and the best attack on a problem may involve finding sub-problems, where "best" may be motivated by computational, inferential, or interpretational reasons. The discovery of such sub-problems might itself be an inferential problem. On the other hand, data often provide partial views of a problem, and the solution may involve fusing multiple data sources. These perspectives of segmentation versus fusion will often not be in conflict, but substantial thought and domain knowledge may be required to reveal the appropriate combination.

One might hope that general, standardized procedures might emerge that can be used as a default for any massive data set, in much the way that the Fast Fourier Transform is a default procedure in classical signal processing. The committee is pessimistic that such procedures exist in general. To take a somewhat fanciful example that makes the point, consider a proposal that all textual data sets should be subject to spelling correction as a preprocessing step. Now suppose that an educational researcher wishes to investigate whether certain changes in the curricula in elementary schools in some state lead to improvements in spelling. Short of designing a standardized test that may be difficult and costly to implement, the researcher might be able to use a data set such as the ensemble of queries to a search engine before and after the curriculum change was implemented. For such a researcher, it is exactly the pattern of misspellings that is the focus of inference, and a preprocessor that corrects spelling mistakes is an undesirable step that selectively removes the data of interest.

Nevertheless, some useful general procedures and pipelines will surely emerge; indeed, one of the goals of this report is to suggest approaches for designing such procedures. But the committee emphasizes the need for flexibility and for tools that are sensitive to the overall goals of an analysis. Massive data analysis cannot, in general, be reduced to turnkey procedures that consumers can use without thought. Rather, as with any engineering discipline, the design of a system for massive data analysis will require engineering skill and judgment. Moreover, deployment of such a system will require modeling decisions, skill with approximations, attention to diagnostics, and robustness. As much as the committee expects to see the emergence of new software and hardware platforms geared to massive data analysis, it also expects to see the emergence of a new class of engineers whose skill is the management of such platforms in the context of the solution of real-world problems.

Finally, it is noted that this report does not attempt to define "massive data." This is, in part, because any definition is likely to be so context-dependent as to be of little general value. But the major reason for sidestepping an attempt at a definition is that the committee views the underlying intellectual issue to be that of finding general laws that are applicable at a variety of scales, or ideally, that are scale-free. Data sets will continue to grow in size over the coming decades, and computers will grow more powerful, but there should exist underlying principles that link measures of inferential accuracy with intrinsic characteristics of the data-generating process and with computational resources such as time, space, and energy. Perhaps these principles can be uncovered once and for all, such that each successive generation of researchers does not need to reconsider the massive data problem afresh.

Appendixes

A

Acronyms

ABC   approximate Bayesian computation
AMT   Amazon Mechanical Turk
API   application programming interface

CCD   charge-coupled device
CPU   central processing unit
CQL   Contextual Query Language

DAG   directed-acyclic-graph
DBMS   database management system
DNA   deoxyribonucleic acid
DSMS   data stream management system

FDR   false discovery rate
FFTW   Fastest Fourier Transform in the West
FPGA   field programmable gate array
FTRL   follow the regularized leader

GB   gigabyte
GFS   Google's File System
GPU   graphics processing unit

HDFS   Hadoop distributed file system

I/O   input/output

| LAPACK | Linear Algebra PACKage library |
| LP | linear programming |

| MCMC | Markov Chain Monte Carlo |
| MLE | maximum likelihood estimation |

| NP | non-deterministic polynomial time |

| OCR | optical character recognition |

| PB | petabyte |
| PCA | principal components analysis |

| QFS | Quick File System |
| QP | quadratic programming |

| RAM | random access memory |
| RDS | respondent-driven sampling |

| S3 | Simple Storage Service |
| SQL | Structured Query Language |

| TB | terabyte |
| TCAM | ternary content addressable memory |

B

Biographical Sketches of
Committee Members

MICHAEL I. JORDAN, *Chair,* is the Pehong Chen Distinguished Professor in the Department of Electrical Engineering and Department of Statistics at the University of California, Berkeley. He received a B.S. in psychology in 1978 from Louisiana State University, an M.S. in mathematics in 1980 from Arizona State University, and a Ph.D. in cognitive science in 1985 from the University of California, San Diego. His research interests are in the field of statistical machine learning, a field that bridges computation and statistics, with ties to information theory, signal processing, algorithms, control theory and optimization theory. One area of his research focus has been probabilistic graphical models, which blends probability theory and graph theory to represent uncertainty on interdependent collections of random variables. He developed new graphical model architectures that have had impact in various applied fields, including bioinformatics, computational vision, speech, natural language processing and information retrieval, and has contributed to the development of a new framework for inference in graphical models based on variational representations of probability distributions. Another area of focus has been nonparametric inference, including both Bayesian nonparametrics, where he developed new models based on the area of stochastic processes known as completely random measures, and frequentist nonparametrics, where he focused on kernel machines, spectral methods, dimension reduction and classification. He has also been interested in developing applications of machine learning to problems in distributed computer systems. In 2010, Dr. Jordan was elected to both the National Academy of Sciences and the National Academy of Engineering.

KATHLEEN M. CARLEY is a professor in the School of Computer Science at Carnegie Mellon University (CMU). She is the director of the Center for Computational Analysis of Social and Organizational Systems, a university-wide interdisciplinary center that brings together network analysis, computer science and organization science and has an associated National Science Foundation (NSF)-funded training program for Ph.D. students. Her research combines cognitive science, social networks, and computer science to address complex social and organizational problems. Her specific research areas are dynamic network analysis, computational social and organization theory, adaptation and evolution, text mining, and the impact of telecommunication technologies and policy on communication, information diffusion, disease contagion, and response within and among groups, particularly in disaster or crisis situations. She and her team have developed infrastructure tools for analyzing large-scale dynamic networks and various multi-agent simulation systems. The infrastructure tools include the ORA, a statistical toolkit for analyzing and visualizing multi-dimensional networks. Another tool is AutoMap, a text-mining system for extracting semantic networks from texts and then cross-classifying them using an organizational ontology into the underlying social, knowledge, resource, and task networks. She is the founding co-editor of *Computational Organization Theory* and has co-edited several books in the computational organizations and dynamic network area.

RONALD R. COIFMAN is a professor of mathematics and computer science at Yale University. His research interests include nonlinear Fourier analysis, wavelet theory, singular integrals, numerical analysis and scattering theory, real and complex analysis, and new mathematical tools for efficient computation and transcriptions of physical data, with applications to numerical analysis, feature extraction recognition, and de-noising. He is currently developing analysis tools for spectrometric diagnostics and hyperspectral imaging. Dr. Coifman is a member of the American Academy of Arts and Sciences and the National Academy of Sciences. He is a recipient of the 1996 DARPA Sustained Excellence Award, the 1996 Connecticut Science Medal, the 1999 Pioneer Award of the International Society for Industrial and Applied Science, and the 1999 National Medal of Science.

DANIEL J. CRICHTON is a principal computer scientist and program manager for the Earth Science Data System and Technology Directorate and the Solar System Exploration Directorate at NASA's Jet Propulsion Laboratory (JPL), where he provides leadership in the development of large-scale, scientific data systems for planetary, Earth, and other data-intensive technology projects. He has served in numerous roles including as principal investigator supporting the research and implementation of new novel approaches for dealing with the capture, management, distribution, and analysis of massive scientific data. He conceived of and built an open-source software framework to enable large-scale data management and sharing of scientific data across organizations that has been accepted into the Apache Software Foundation. He has served on a number of committees for NASA, the National Institutes of Health, and other agencies. He has authored more than 100 book chapters and papers on the topic of data-intensive systems. He has a B.S. in information and computer science from the University of California, Irvine, and an M.S. in computer science from the University of Southern California.

MICHAEL J. FRANKLIN is the Thomas M. Siebel Professor of Computer Science at the University of California, Berkeley, specializing in large-scale data management applications and infrastructure. Dr. Franklin is a member of the Database and Operating Systems and Networking Technology groups. He is director of the Algorithms, Machines, and People Laboratory (AMPLab), where he collaborates with students, postdoctoral researchers, and faculty who specialize in cloud computing, statistical machine learning, networking, and other important topics necessary for building scalable data-intensive systems. He is a co-founder of Truviso, a high-performance analytics software company in Foster City, California.

ANNA C. GILBERT is a professor in the Department of Mathematics at the University of Michigan. She has an S.B. degree from the University of Chicago and a Ph.D. from Princeton University, both in mathematics. In 1997 Dr. Gilbert was a postdoctoral fellow at Yale University. From 1998 to 2004 she was a member of technical staff at AT&T Labs-Research in Florham Park, New Jersey. Her research interests include analysis, probability, networking, and algorithms, and she is especially interested in randomized algorithms with applications to harmonic analysis, signal and image processing, networking, and massive data sets.

ALEX GRAY is director of the Fundamental Algorithmic and Statistical Tools Laboratory (FASTlab) at the Georgia Institute of Technology. Dr. Gray received bachelor's degrees in applied mathematics and computer science from the University of California, Berkeley, and a Ph.D. in computer science from Carnegie Mellon University. He worked in the Machine Learning Systems Group of NASA's JPL for 6 years. FASTlab works on the problem of how to perform machine learning/data mining/statistics on massive data sets and related problems in scientific computing and applied mathematics. Employing a multidisciplinary array of technical ideas (from machine learning, nonparametric statistics, convex optimization, linear algebra, discrete algorithms and data structures, computational geometry, computational physics, Monte Carlo methods, distributed computing, and automated theorem proving), his laboratory has developed the current fastest algorithms for several fundamental statistical methods, and is in the process of developing new machine learning methods for difficult aspects of real-world data, such as in astrophysics and biology. This work has enabled high-profile scientific results that have been featured in *Science* and *Nature*. Dr. Gray has received an NSF CAREER award, two best-paper awards, and two best-paper award nominations.

TREVOR HASTIE is a professor in the Department of Statistics at Stanford University and the Division of Biostatistics of the Health, Research, and Policy Department in the Stanford School of Medicine. His main research contributions have been in the field of applied nonparametric regression and classification, and he has co-written two books in this area, *Generalized Additive Models* and *Elements of Statistical Learning*. He has also made contributions to statistical computing, co-editing a large software library on modeling tools in the S language (*Statistical Models in S*, 1992), which form the basis for much of the statistical modeling in R and S-plus. His current research focuses on applied problems in biology and genomics, medicine, and industry, in particular data mining, prediction, and classification problems.

PIOTR INDYK is a professor in the Department of Electrical Engineering and Computer Science at Massachusetts Institute of Technology (MIT). He joined MIT in September 2000 after earning a Ph.D. from Stanford University. Earlier, he received a magister degree from Uniwersytet Warszawski, Poland, in 1995. Dr. Indyk's research interests include computational geometry (especially in high-dimensional spaces), algorithms using sublinear time and/or space, and streaming algorithms. He is also interested in algorithmic coding theory and pattern-matching problems.

THEODORE JOHNSON is a research scientist in the Database Research Department at AT&T Labs-Research. He received a B.S. in mathematics from the Johns Hopkins University in 1986 and a Ph.D. in computer science from the Courant Institute of New York University in 1990. From 1990 through 1996, he was an assistant professor, and then an associate professor, in the Computer and Information Science and Engineering Department at the University of Florida. In 2004 he received an AT&T Science and Technology Award for his work in the Bellman database browser, and in 2010 he was made an AT&T fellow. He has co-authored two books, *Distributed Operating Systems and Algorithms* and *Exploratory Data Mining and Data Cleaning*.

DIANE LAMBERT is a research scientist at Google, Inc. She has previously served as head of statistics and data mining research at Bell Laboratories from 1997 to 2005 and was a member of its technical staff from 1994 to 1997. She was a tenured member of the faculty at CMU from 1980 to 1986 and was also a visiting associate professor at University of Chicago from 1984 to 1986. She has held numerous editorial and program committee positions.

DAVID MADIGAN is a professor of statistics at Columbia University. He received a B.A. in mathematical sciences (1984) and Ph.D. (1990) in statistics from Trinity College in Dublin, Ireland. He was previously the dean of physical and mathematical sciences at Rutgers, The State University of New Jersey. He has received numerous honors that include the Institute of Mathematical Statistics Medallion Lecturer, fellow of the Institute of Mathematical Statistics, and being named as the "36th Most Cited Mathematician in the World, 1995-2005."

MICHAEL MAHONEY is an engineering research associate in the Department of Mathematics at Stanford University. His research interests are algorithmic and statistical aspects of modern large-scale data analysis; design and analysis of algorithms for matrix, graph, and regression problems; statistical data analysis in large-scale scientific and Internet applications; applications to the analysis of large social and information networks; applications to DNA microarray and single nucleotide polymorphism data; and randomized algorithms for large linear algebra problems. Much of his current research focuses on geometric network analysis, developing approximate computation and regularization methods for large informatics graphs; applications in large social and information networks; and statistical data analysis of extremely large data sets. Recently, this work led to improved algorithms for two classical linear algebra problems.

F. MILLER MALEY is a researcher on the staff at the Communications Research Center (CRC), Princeton, a division of the Institute for Defense Analysis that supports National Security Agency research interests. He is also co-chair of the CRC's SCAMP program on supercomputing. He was a visiting research fellow at Princeton University from 1987 to 1990. Dr. Maley received a B.S. in mathematics and physics from Amherst College in 1983, a Ph.D. in computer science from MIT in 1987, and a Ph.D. in mathematics from Princeton University in 1996. He is the author or co-author of 17 classified papers. His awards include the NSF Mathematical Sciences Postdoctoral Research Fellowship (1987-1990) and Office of Naval Research Graduate Fellowship (1983-1987).

CHRISTOPHER OLSTON is a staff research scientist at Google, Inc. Previously, he was a principal research scientist at Yahoo! Research. His research interest is data management, focusing especially on Web data management challenges. Dr. Olston received his Ph.D. in computer science in 2003 from Stanford University, supported by fellowships from the university and NSF. He received his bachelor's degree in electrical engineering and computer sciences from the University of California, Berkeley, with highest honors. He has previously held teaching and research positions at Yahoo! Research, Carnegie Mellon University, Stanford University, Xerox Palo Alto Research Center, the University of California, Berkeley, and Informix Software, Inc.

YORAM SINGER is a senior research scientist at Google, Inc. Before joining Google, he was an associate professor at the School of Computer Science and Engineering of Hebrew University of Jerusalem, and before that he was a member of the technical staff at AT&T-Research. Dr. Singer received his B.Sc. and M.Sc. degrees in computer science from the Technion and his Ph.D. in computer science from Hebrew University.

ALEXANDER SANDOR SZALAY is a professor in the Department of Physics and Astronomy of Johns Hopkins University. His research interests are theoretical astrophysics and galaxy formation. His research interests are multicolor properties of galaxies, galaxy evolution, the large-scale power spectrum of fluctuations, gravitational lensing, pattern recognition and classification problems, the SDSS project, and large scalable databases. He is a leader in the use of massive data as input for scientific research.

TONG ZHANG is a professor of statistics at Rutgers University. His research interests are machine learning, statistical and numerical computation, and design and theoretical analysis of statistical algorithms. He has worked extensively in large-scale data analysis and statistical modeling, especially in text mining, natural language processing, search, and various other Web applications. Dr. Zhang received a Ph.D. in computer science from Stanford University in 1998. After graduation, he worked at IBM T.J. Watson Research Center and then Yahoo! Research in New York.